The Ethics of Military Privatization

This book explores the ethical implications of using armed contractors, taking a consequentialist approach to this multidisciplinary debate.

While privatization is not a new concept for the US military, the public debate on military privatization is limited to legal, financial, and pragmatic concerns. A critical assessment of the ethical dimensions of military privatization in general is missing. More specifically, in light of the increased reliance upon armed contractors, it must be asked whether it is morally permissible for governments to employ them at all. To this end, this book explores four areas that highlight the ethical implications of using armed contractors: how armed contractors are distinct from soldiers and mercenaries; the commodification of force; the belligerent equality of combatants; and the impact of armed contractors on the professional military. While some take an absolutist position, wanting to bar the use of private military altogether, this book reveals how these absolutist arguments are problematic and highlights that there are circumstances where turning to private force may be the only option. Recognising that outsourcing force will continue, this book thus proposes some changes to account for the problems of commodification, belligerent equality, and the challenge to the military profession.

This book will be of interest to students of private security, military studies, ethics, security studies, and IR in general.

David M. Barnes is a career US Army officer and Assistant Professor at the United States Military Academy. He has a PhD in Philosophy from the University of Colorado, USA.

Military and Defence Ethics
Series Editors

Don Carrick – Project Director of the Military Ethics Education Network based in the Institute of Applied Ethics at the University of Hull, UK.

James Connelly – Professor of Politics and International Studies, Director of the Institute of Applied Ethics, and Project Leader of the Military Ethics Education Network at the University of Hull, UK.

Paul Robinson – Professor in Public and International Affairs at the University of Ottawa, Canada.

George Lucas – Professor of Philosophy and Director of Navy and National Programs in the Stockdale Center for Ethical Leadership at the U.S. Naval Academy, Annapolis MD, USA.

There is an urgent and growing need for all those involved in matters of national defence – from policy makers to armaments manufacturers to members of the armed forces – to behave, and to be seen to behave, ethically. The ethical dimensions of making decisions and taking action in the defence arena are the subject of intense and ongoing media interest and public scrutiny. It is vital that all those involved be given the benefit of the finest possible advice and support. Such advice is best sought from those who have great practical experience or theoretical wisdom (or both) in their particular field and publication of their work in this series will ensure that it is readily accessible to all who need it.

Also in the series

The Ethics of Military Privatization

The US armed contractor phenomenon

David M. Barnes

Routledge
Taylor & Francis Group

LONDON AND NEW YORK

First published 2017
by Routledge
2 Park Square, Milton Park, Abingdon, Oxon OX14 4RN

and Routledge
605 Third Avenue, New York, NY 10017

First issued in paperback 2021

Routledge is an imprint of the Taylor & Francis Group, an informa business

British Library Cataloguing in Publication Data
A catalogue record for this book is available from the British Library

Library of Congress Cataloging in Publication Data
Names: Barnes, David M., COL.
Title: The ethics of military privatization: the US armed contractor phenomenon/David M. Barnes.
Other titles: US armed contractor phenomenon
Description: New York, NY: Routledge, [2016] | Includes bibliographical references and index.
Identifiers: LCCN 2016010842 | ISBN 9781472464439 (hardback) | ISBN 9781315572444 (ebook)
Subjects: LCSH: Private military companies–United States–Moral and ethical aspects. | Private security services–United States–Moral and ethical aspects. | Contracting out–United States–Moral and ethical aspects. | Privatization–United States–Moral and ethical aspects. | War–Moral and ethical aspects.
Classification: LCC UB149 .B37 2016 | DDC 172/.42–dc23
LC record available at http://lccn.loc.gov/2016010842

ISBN 13: 978-0-367-78727-1 (pbk)
ISBN 13: 978-1-4724-6443-9 (hbk)

Typeset in Times New Roman
by Sunrise Setting Ltd., Brixham, UK

For "Little Lucky" Kate—my ski buddy, hiking buddy, and partner in storytelling.

For "Lucky" Jack—my right hand man, sports and adventure explorer.

For Alli: you are my everything

Contents

Acknowledgments

Wrestling with difficult topics such as military privatization requires a deliberate process, and this one is literally years in the making. Far too quickly, I think, we look for a solution to a problem, when we should be spending some time, perhaps most of the time, trying to sort out what it is we are trying to solve. The more I researched and thought about the armed contractor phenomenon, I've come to realize that we have moved to accept privatization as part of how we fight and are now really beginning to question how we got to this point. I think that the debate over privatization, in particular the employment of armed contractors, has improved, but I also think that we generally assume that privatization is here to stay. So, while the public and academic debate focuses on better oversight and control, I think we should also ask the normative question: whether we should outsource certain services, and what are the implications of doing so. I have been fortunate and been able to take a step back and think through the history and employment of armed contractors, and I wished I had understood how and in what way certain services were privatized earlier in my career.

No work like this is completed in isolation, and I am indebted to many over the years who have both shaped my operational experience and educational development, and more recently have encouraged me to complete this project through moves, different positions, even deployments. I have been fortunate to work with the best soldiers, noncommissioned officers, and officers for more than twenty five years. They molded me, improved my thinking, and challenged me in ways I could not have imagined in my days as a cadet. I've had the privilege of leading the finest and been led by the finest, and I will always be grateful. I have also met and worked with some amazing contractors, and while some may disagree with my findings, I know we share a commitment to improving our armed forces and its privatized support. To the friends, colleagues, former student, and soldiers lost, I am honored to have served with you.

I approached Alastair Norcross with my crazy idea to look at armed contractors through an ethical lens, and he gladly agreed to help in my process. Together with David Boonin, Ben Hale, Eric Chwang, Chris Heathwood, and colleagues from the University of Colorado, Boulder, Alastair challenged my arguments, made suggestions for improvement, and helped me frame the initial discussion. I was also able to present portions of this work in a number of conferences. I would like to thank members of the audience, especially Fernando Teson and James Pattison, at the 2010 International Studies Association (ISA), as well as Sebastian Schilling, Bernhard Koch, and other participants of the 2010 Zentrum für ethische Bildung in den Streitkräften (zebis) conference. Each question and discussion sharpened my ideas and helped make this book better.

Speaking of conferences, I cut my teeth in military ethics as a long time participant in the International Society for Military Ethics (ISME). Formally known as the Joint Services Conference on Professional Ethics—JSCOPE, ISME's meetings have been held each year since 1979, and I have been fortunate to both present and be a member of the audience in many of these sessions. As a member of ISME, I have been fortunate to work with some outstanding professionals who are truly dedicated to furthering the vital discussions and education of the ethics of war. My ISME colleagues were and are a primary source of inspiration, mentorship, and friendship for me. In particular, I thank Richard Schoonhoven, Martin Cook, Jim Cook, George Lucas, Pauline Shanks-Kaurin, Rebecca Johnson, Stéphanie Bélanger, and many others who continually motivated me to carry on with this project.

I also want to thank my colleagues at the United States Military Academy. I am indeed fortunate to work with such a fantastic faculty and staff. Working with them and teaching the next generation of Army leaders is truly an honor and highlight of my career. I am grateful that Tony Hartle, Rick Kerin, Scott Krawczyk, and Dan Zupan were willing to take a chance and allow this tanker to develop as a philosopher.

I cannot thank my family enough, and it has been a family project. Mom and dad helped edit earlier drafts, and Jack, Kate, and Allison have endured too many moves to count, numerous odd hours of reading and writing, and several deployments. They have stood by me, inspired me, and humbled me. You make me proud and make me better.

Abbreviations

AO	Area of Operations
AWE	Army Warfighting Experiment
CGSC	Command and General Staff College
CIA	Central Intelligence Agency
CPA	Coalition Provisional Authority (Iraq)
DDE	Doctrine of Double Effect
ECP	Entry Control Point
EO	Executive Outcomes
EU	European Union
GAO	Government Accounting Office
ILE	Intermediate-Level Education
IRS	Internal Revenue Service
ISIS	Islamic State
ISOA	International Stability Operations Association
ITAR	International Traffic Arms Regulations
JSOC	Joint Special Operations Command
JWT	Just War Tradition
LOAC	Law of Armed Conflict
LOGCAP	Logistics Civil Augmentation Program
MEC	Moral Equality of Combatants
MEJA	Military Extraterritorial Jurisdiction Act
MNC-I	Multi-National Corps-Iraq
MPRI	Military Professional Resources Inc.
NDAA	National Defense Authorization Act
NGOs	Nongovernmental Organizations
OAS	Organization of American States
OCS	Officer Candidate School
OSCE	Organization for Security and Cooperation in Europe
PMC	Private Military Company
PMSC	Private Maritime Security Company
PNG	Papua New Guinea
POW	Prisoner of War
PSCs	Private Security Companies

PSD	Personal Security Detail
PTSD	Post-traumatic Stress Disorder
ROTC	Reserve Officers' Training Corps
RSB	Russian Security Systems Group
RUF	Revolutionary United Front (Sierra Leone)
SAS	British Special Air Service
SAW	Squad Automatic Weapon
SCR	Security Council Resolution
SOFA	Status of Forces Agreement
SOP	Standing Operating Procedure
SOSi	SOS International
TIS	Total Intelligence Solutions
TSA	Transportation Security Administration
UCMJ	Uniform Code of Military Justice
VBIED	Vehicle Borne Improvised Explosive Device
VPD	Vessel Protection Detachment

1 Introduction

Close combat privatization

> ... I also believe that it is important actually to make the argument against mercenaries—and not merely to assume it.
>
> (Walzer 2008)

> To put it in another way, the war in Iraq would not be possible without private military contractors. This is critically important. Contrary to conspiracy theories, the private military industry is not the so-called "decider," plotting out wars behind the scenes like Manchurian Global. But it has become the ultimate enabler, allowing operations to happen that might be otherwise politically impossible.
>
> (Singer 2007a)

America woke on the morning of 1 April 2004 to the headlines "4 From US Killed in Ambush in Iraq; Mob Drags Bodies" and "Four Civilians Killed in Iraq Worked for NC Security Firm."[1] At first glance, the headlines seemed to be a repetition of the same old story: four more casualties in Iraq. That four civilians were killed in Fallujah was not much of a surprise; there had already been many civilian and military casualties in Iraq. What did cause some to pause, however, was the manner in which they were killed and how their bodies were burned and displayed on international television, like some macabre war trophy. The United States (US) population began to become more interested; the inhuman treatment of these men's remains generated questions: What were these four civilians doing in a known combat zone, relatively unprotected and unsupported? Where was the military?

Of course, these were not your average men from the street; they were seasoned former special operations soldiers, armed and escorting a convoy.[2] People began to wonder about the identity of these armed civilians. If they were not soldiers or Central Intelligence Agency (CIA) operatives, then who were they? What if these same four men had been captured instead of being killed? What if the enemy, or a government that did not support the Iraqi invasion, decided to prosecute them as illegal combatants? Unfortunately, there was and currently is no universally agreed upon standard for these privatized, armed employees. Their legal status remains underdefined.[3]

Privatization (or outsourcing or contracting) is not a new concept to the US military. As far back as the American Revolutionary War, Washington hired long oarsman from Massachusetts to move his army by water, including the famous crossing of the Delaware River, and German soldier of fortune, Baron von Steuben, was indispensable in helping transform Washington's army.[4] This tradition of military privatization continues today where the military has outsourced many of its requirements, especially in logistics. In the last few decades, however, the US military has grown increasingly reliant on privatization. From deployments to Southwest Asia, Afghanistan, and the Balkans to daily operations in home bases, every aspect of the military seems to have a privatized component. Company names such as KBR, DynCorp, Vinnell, MPRI, Cubic, and Northrop Grumman are all common place in military vernacular. Today, the name "Blackwater" invokes strong emotional reactions both here and abroad. And, while the wars in Iraq and Afghanistan seem to be winding down, the continued turmoil in these places, as well as in other areas, such as Libya, Syria, and Yemen, belie the perception that there is a lessening of reliance upon military outsourcing—an outsourcing dividend.

Even portions of post security are privatized; drive to the gate at almost any post or base and look at the logos on the gate guards. But, what if there was an incident at the post gate one morning, an incident where one of these civilian guards would have to use deadly force? Now consider the same incident in some foreign land—Iraq for instance (or Afghanistan or even Somalia)—a land where the United States does not have a Status of Forces Agreement (SOFA) or a place without a strong central government. It becomes increasingly difficult to legally categorize these armed contractors, and what they can and cannot do.

The good news is that the use of armed contractors from privatized military firms has entered public debate. Even Congress has begun to address the phenomenon of contracting force.[5] Discussion over the employment of these private forces is also developing in academic circles. Unfortunately, while it is positive to note that the debate on military privatization continues, this debate is limited, focused on legal, financial, and pragmatic concerns, such as oversight, which are covered in P. W. Singer's *Corporate Warriors* (2003) and D. Avant's *The Market for Force* (2005). In his *Private Sector, Public Wars*, James Jay Carafano (2008) argues for their employment, but he bases his case on the assumption that this privatized force is an evolution of warfare. There has even been the launch of an effort to consider these questions from an ethical perspective.[6] But, while I think that these are good contributions to the debate, they fall short. What is missing from this debate is a critical assessment of the ethical dimensions of military privatization in general; more specifically, in light of the profusion of these contracts, it must be asked whether it is morally permissible for governments to employ armed military contractors. This is what this project intends to do.

The astute reader will recognize that many of the mainstream arguments over military privatization, as well as the examples I will discuss, are clearly legal concerns. Some theorists, myself included, believe that international law tracks morality in terms of the Just War Theory; others disagree, at least to an extent. So, while the questions about the status of the Blackwater contractors killed in Fallujah, for example, clearly illustrate legal concerns such as the question of combatancy under the law of armed conflict (LOAC), I think that their status, as well as what happened to them, is also up for ethical discussion. This book, then, is not meant to provide a thorough legal exploration of military privatization and the use of armed contractors, but I will use the relevant legal concerns to inform the moral discussion, and I will point out those areas where the law and the ethical conclusions seem to diverge.

How is this book different?

The genesis of this book and my relationship with military privatization really began on the morning of 6 May, 2004. My morning routine in Iraq was violently interrupted by a car bombing: a Vehicle Borne Improvised Explosive Device (VBIED) detonated at a checkpoint outside the bridge over the Tigris River in central Baghdad. "Seven killed in Baghdad suicide blast," read *The Guardian's* (2004) headline, but for me it was much more up close and personal. There were certainly more deadly attacks during that tour in Iraq and, hence, only one US soldier died and another 25 people were injured that morning. But, this attack set the tone for my unit for the rest of that deployment, and I discovered that the US military and government agencies were not the only players in the complex post invasion Iraqi environment. It was this incident, and the variety of different entities who responded, that opened my eyes to how much more my military and government had grown to rely upon privatization. And, the US government is not the only employer. My interest in military privatization partly grew from wholly pragmatic concerns: since we controlled access into and out of the International Zone (sometimes colloquially called the "Green Zone") and the territory immediate outside the walls, we needed to know the tenants and the visitors who frequented our area of operations. I had worked with contractors in other places in other capacities before—even in Afghanistan. But, the nature of this mission in Iraq was such that I would come to work with them (even around them) on a regular basis. Then, over time, especially after the deployment, I began to wonder how these private companies had become so important, almost as Singer notes, ubiquitous to military operations. Being a philosopher, I also began to ask the question: "Why?"

To better understand my motivation for this project and to frame it in such a way, I must digress for a minute and provide a bit of my background, for it is my background and my experiences that, I think, provide a unique perspective in the military privatization debate: one that has so far been missing.

I am a career soldier: a US Army officer. Having graduated from West Point, I have served as an Armor officer for over 25 years. I have been stationed overseas, and I have been on multiple deployments, including Iraq and Afghanistan. As I mentioned, soldiers from my unit were the ones who responded to the VBIED on the bridge, and soldiers from my battalion fought in Fallujah after the Blackwater (now Academi) contractors were killed in 2004. My units have suffered casualties, yet so too have contractors been targeted and killed. I have worked hand in hand with many contractors, from many different companies, and together we have seen our share of combat. I have witnessed examples of true professionalism and exemplary courage of these same contractors. Thus, I do not question the use of military contractors lightly.

But, I also have had a unique career as an academic and as a combat arms officer; I have had a boot in both camps. I am one of the few, and perhaps the senior, active serving Army officer with a PhD in Philosophy. While I am interested in the philosophy of mind, political philosophy, and ethics, my specialty (perhaps to no surprise) is military ethics. I not only have served, and am currently serving, on the senior faculty at West Point, but I have written and presented at a number of internationally recognized academic forums. And, I have not abandoned one camp for the other. I returned from another deployment recently, and while these two career paths seem totally distinct, and their differing views intractable,[7] they are complimentary in a number of ways. I hope that Thucydides would not consider me a fool or a coward, but I also do not pretend to offer the definitive answer. Rather, what I bring is a capacity to look at military problems, like privatization and its effect of the military as a profession, with an eye to the underlying theory—theory that I have been able to put into practice.

Some may want to take an absolutist position, barring the use of private military altogether. While these types of arguments may satisfy many opponents' intuitions against privatized force, these absolutist arguments are problematic. Furthermore, there seems to be those circumstances, perhaps ones so dire, that turning to private force may be the only option. Thus, as will be clear to the reader, I take a consequentialist approach to this multidisciplinary armed contractor debate, one informed by contemporary Just War Theory.

Overall, this book is organized into seven chapters, outlined as follows:

1   Introduction
2   The Armed Contractor Phenomenon—A Contemporary Debate with a Long History
3   Mercenaries, Soldiers, and Armed Contractors—An Explication
4   Armed Military Privatization and the Commodification of Force
5   The Belligerent Equality of Armed Contractors?
6   The Challenge of Military Privatization to the Military Profession
7   The "Second Contractor War" and the Future of Armed Contractors

Chapter 2 sets the stage. The story must start somewhere, and the contemporary story of armed contractors and military privatization in general begins long ago. Here, I briefly explore the contemporary armed contractor phenomenon and introduce some of the moral complications that their employment raises. Then, I describe in greater detail the history behind the armed contractor phenomenon. From mercenaries to privateers to the groups like Academi (formerly Blackwater and Xe), G4S, and Triple Canopy today, I establish the background and lineages of these private military corporations and their relationship to the states that hire them. Those who are already very familiar with the debate might question this thorough historical approach. Nevertheless, apart from a valuable academic exercise, this review is critical for assessing how the military, and these contractors and their companies share the battlefield and often the mission. Furthermore, I was asked on a number of occasions to explain who these companies were and why there were there. I tried to explain military privatization as best I could, but we were not trained to integrate with them to such a high degree. Doing this historical analysis also helps lend shape to defining the overall ethical problem and why the armed contractor phenomenon is morally problematic. Identifying the current issues and historical context for the use of armed contractors provides a common backdrop to explore the phenomenon and its contemporary debate. And, some case studies drawn from my and other's experiences will help illustrate.

In Chapter 3, I explore some critical concepts and terms that inform the current project, such as how one defines who are armed contractors. While we had trained for "civilians on the battlefields" or "contractor on the battlefields" ("COBs" for short), especially after deployments to Bosnia and Kosovo, we were not prepared to fully grasp how many different private companies would be working with us and near us. Why, for example, a soldier would ask, does this contractor make ten times what he did for the seemingly same job? "Isn't she a mercenary?" Therefore, I investigate the question of whether the armed contractors of today are, as their proponents argue, vastly different from mercenaries. Some have argued that the distinctions are too fine; they stem from trying to avoid the intuitive negative reaction when using a loaded term, such as "mercenary."[8] Others, such as Blackwater International founder Erik Prince, think that it is this distinction itself that makes private security companies like his legal corporations—ones not subject to anti-mercenary laws.[9] Private security employees, some may argue, can use force, while mercenaries must not. I also attempt to locate and describe the presumed differences between armed contractors and professional soldiers in order to lend substance to categorizing terms that are loosely used by both the media and theorists alike, where none of these seem to sufficiently clarify their use and relationship to the military on the ground. This chapter therefore establishes a base-line for these terms as well as providing background to the problem, which are necessary and may go far to illuminate the armed contractor debate. These definitions may point to ways to better incorporate these companies into future operations.

The chapter on the commodification of force (Chapter 4) looks at whether force itself is in fact a commodity; since these military companies are financially contracted, it seems true. Then, I explore the normative question, "Should force be a commodity?" I take Max Weber's notion that the state (and only the state) has the monopoly on the legitimate use of force, and that the contracting of private security companies are incorrectly viewed by their proponents as a natural evolution of force as a commodity. I demonstrate that the armed contractor phenomenon itself commodifies force, and it does so in a distinct manner, different from and detrimental to the state's purview to raise and maintain a military. Because force should not be a commodity and states have a monopoly on force, then even though states can in a sense delegate the use of force, say by hiring armed contractors, baring extreme circumstances, they should not do so.

In Chapter 5, "The Belligerent Equality of Armed Contractors?," I argue against Jeff McMahan's (2006c) revisionist rejection of the Just War Theory of the moral equality of combatants (MEC). Although he raises some strong problems that stem from MEC, I argue that MEC morally grounds how just wars are fought, and it is the concept of MEC itself that demonstrates that armed contractors are not the legal or moral equivalent of soldiers, nor should they always be considered civilians. Accepting the revisionist reasoning raises several concerns, foremost defining who combatants are. The question of determining not just the legal combatant status but also the moral combatant status plays an important role in the jus in bello tenet of discrimination, and it will show that while armed contractors are acting as combatants (they do not appear to be non-combatants), they are not combatants with the same moral status of soldiers. Moreover, this moral distinction extends beyond mere legality or McMahan's just-unjust distinction. There exists a MEC between soldiers that does not extend to armed contractors.

After addressing Commodification and Belligerent Equality, this project turns to the argument against employing armed contractors based upon their incompatibility with and weakening of the state's professional military. In "The Challenge of Military Privatization to the Military Profession," I outline the concept commonly held by the US military that the military itself is a profession—a profession of arms, and the soldiers, sailors, marines, and so on are members of this profession. That the military is a profession has been argued elsewhere (e.g. Samuel Huntington's *The Soldier and the State* (1957)).[10] Thus, I will focus on the challenges that armed contractors bring to the military as a profession, how these challenges potentially usurp the military profession's authority, and why this erosion of the authority of the military profession negatively affects public policy, including the jus ad bellum decision to go to war.[11]

These arguments together demonstrate that refraining from using armed contractors results, all things considered equally, in a better state of affairs. States should not normally revert to employing armed contractors. As I will show, the costs, financial and otherwise, outweigh the benefits of their employment. Only in extreme circumstances, where the benefits of their use counterbalance the costs, would armed contractors become a morally viable option. I will also touch upon how these arguments affect the privatization of other functions within the military (e.g. logistics) and explore the possible ramifications for proponents of privatization in general. Additionally, I offer some observations on the current employment of armed contractors, as well as implications for their further employment. Here I recognize that, pragmatically, outsourcing force will continue; therefore, I propose some changes to account for the problems of commodification, belligerent equality, and the challenge to the military profession.

Finally, I will discuss the current state of military privatization and will ask what the future might have in store. For example, Ann Hagedorn (2014) asks if America's Second Contractors' War is drawing near? In 2010, President Barack Obama declared the end of the Iraq war, and much of that fight and our current involvement in the Middle East are carried out by a privatized military. Some even called the post 9/11 Operations Iraqi Freedom and Enduring Freedom the "First Contractors' War."[12]

At the same time, with the end of operations in Iraq and Afghanistan, news about private military companies seemed to all but have disappeared. Nevertheless, privatized security remains a seemingly essential part of American operations.

For example, with the Islamic State (ISIS) attacks in Iraq widening in 2014, seemingly threatening Baghdad itself, the Obama administration announced that 300 advisors would assist the Iraqi military. This has expanded to over 1600, including a division headquarters, as operations expand into Syria (Records 2014). Simultaneously, it was emphasized that there would be "no boots on the ground," a similar mantra echoed during the Syrian Crisis. Yet, a 2014 *New York Times* article (Rubin and Nordland) mentioned that these advisors would be supported by privatized security, specifically, "more than 1,000 American private security guards [will go] to Iraq to protect the 300 military and intelligence advisors." And, this number increased with the additional headquarters and advisor personnel. The administration also asked Congress for $500 million to train and equip moderate Syrian Rebels. Secretary of Defense Hagel ordered detailed Pentagon planning, and much of this planning includes how private military companies will be employed in support of this mission, especially if the "no boots" constraint remains.

At its height, in 2008 there were 242,558 contractors working for US Central Command in the region. By July 2014, that number had shrunk to 66,123 ("Contractor Support" 2014). Yet, the emerging conflicts offer new privatization prospects. Roger Carstens, a former special operations officer who has served as a contracted military advisor in Somalia and Afghanistan, said, "this new war will present an opportunity for the companies that have a resident train and advising capability to contribute to this new effort" (Lake 2014). The $500 million that President Obama asked Congress to authorize to train a new Syrian opposition out of Saudi Arabia would be part of a $5 billion fund Obama requested from Congress to help train and equip US allies to fight terrorists.

Eli Lake (2014) notes that the Pentagon asked military contractors to participate in two important surveys. The first one, issued in July 2014, asked the industry to give a rough cost estimate of building a new network of ten ground based communications stations. The second one was more telling. It asked for estimates of the cost for "Security Assistance Mentors and Advisors" for Iraq's ministry of defense and the Iraqi Counterterrorism Service. Lake writes: "Iraq recently promised immunity for US troops—and it is likely Baghdad will do the same for contractors too. After all, Iraq's government has also formally requested US assistance in fighting ISIS and that help was clearly going to include military contractors."

At the same time, the anti-piracy mission continues for many private military companies, and according to Tim Shorrock (2014), "A massive, $7.2 billion Army intelligence contract signed [12 September 2014] underscores the central role to be played by the NSA and its army of private contractors."[13] Furthermore, the US and her allies are not the only private military players now: Russia, Israel–even China have entered the market for force.

So, even with the drawdown of forces in Iraq and Afghanistan, there is a continuing perceived need for contractors, and this need looks like it will increase. Doug Brooks, the president emeritus of the International Stability Operations Association, a trade association for professional military contractors notes: "There is going to be business, you could say these [contractors] are shoes instead of boots on the ground" (Lake 2014). But, maybe, just maybe, the contractors themselves will provide the boots: Erik Prince himself has argued that we should let contractors fight ISIS (Lamothe 2014).

However, before I proceed to the description and background of the Armed Contractor Phenomenon (Chapter 2), I must briefly discuss two assumptions that inform my project: state sovereignty and Weber's monopoly of force.

The Weberian state

Every state is founded on force,' said Trotsky at Brest-Litovsk. That is indeed right. If no social institutions existed which knew the use of violence, then the concept of 'state' would be eliminated, and a condition would emerge that could be designated as 'anarchy'.... Of course, force is certainly not the normal or the only means of the state—nobody says that—but force is a means specific to the state. Today the relation between the state and violence is an especially intimate one.... Today, however, we have to say that a state is a human community that (successfully) claims the monopoly of the legitimate use of physical force within a given territory.

(Weber 1926)

The debate over the use of private military and security companies is currently being conducted within the context of state-dominated geopolitics. One question that might be interesting, at least to political theorists, is whether the bias against privatized force may lessen or whether the debate over their use becomes irrelevant in a different world order. Although I will discuss some of these possibilities, I focus on the privatized military debate within the current international order. I do so partly because some just could not imagine nor participate in this discussion if it were presented in terms of other possible worlds, and I do so because the conclusions have direct impact on the use of force here and now, in the actual world—the world where our soldiers have deployed and fought.

There are two subjects—really assumptions on my part—that while simply accepted by others, need a bit of further discussion, especially how they relate to the armed contractor phenomenon. The first is state sovereignty, an assumption so widely accepted in international relations that it is not only assumed but is also treated as a given. The second concerns Max Weber's conception that the state has a monopoly on the legitimate use of force. Of course, there is plenty of empirical evidence that entities other than states have used force, thus calling the "monopoly" into question. But, like state sovereignty, I take Weber's claim to be true. So, since others might try to show that my arguments are grounded in these assumptions, I will say a few words about both state sovereignty and the monopoly of force.

A few words about state sovereignty

... sovereignty is the international institution that organizes global politics.

(Thomson 1994)

Of course, I need to be up front and acknowledge that in spite of its wide acceptance, there are certainly difficulties with defining "sovereignty." Nevertheless, debate over state power, authority, and responsibility—especially when the debate involves military maters—assumes that the state is sovereign. The symbol of state sovereignty "is its ability to monopolize the means of violence—in other words, raising, maintaining, and using military forces" (Isenberg 2009). From a states' rights perspective, states enjoy the right to political sovereignty and the right of territorial integrity. States, therefore, raise and maintain armies to deter and protect the state and its citizens against foreign violations of these rights. Setting other, mostly revolutionary, exceptions aside, the evolution of the international system since Westphalia is that wars have been conducted using state-raised forces. I therefore adopt a traditional view of state sovereignty. Of course, states rarely act unilaterally. The European Union (EU) is an example of sovereign member states maintaining their own monopoly on the use of force, while collectively practicing market sovereignty.[14]

One might thus view the recent revival and employment of private military companies, and armed contractors in particular, as one of several types of challenges to the current international norm of state sovereignty. The rise of non-governmental organizations (NGOs) and international organizations such as the EU, the increasing influence of multinational corporations, and the globalization phenomenon all are stretching the traditional view of the role of the state, even what defines the state itself. These challenges can be clearly seen in the developing world where the state is ineffective, even absent. The developing world, to no surprise, has seen the greatest increase in the use of private military companies. Of course the story is more complex. It is naïve to think that only strong states shy away from private force; there are other factors at play. "In particular, the changing nature of state competence, the growth of private sector, market deregularization and military downsizing," Kinsey (2006, 64) argues, "all contributed to the rise of [private military companies] on the international stage."

While privatized force never disappeared, its widespread use by states declined beginning in the thirteenth century, and much of the reliance upon private force was replaced by the practice of using citizen-based military forces in order to conduct legitimate violence. Part of this attempt aimed to reduce the illicit violence of private groups within state boundaries. Part was due to perceived cost savings and increased proficiency on the battlefield. Some was due to a perception that mercenaries and the practice of hiring them was morally corrupt. Privatized force seemed relegated to, and endemic of, the developing world where states were either ineffective or non-existent (similar perhaps to the reliance upon mercantile companies and privateers in the New World). Privatized military remained outside of the public interest through much of the twentieth century. Today, however, private military companies are increasingly employed by strong states in areas outside the developing world, including within their own territory (e.g. New Orleans after Katrina or the Reserve Officer Training Corps (ROTC) training in universities). This is where questions concerning challenges to the conception of sovereignty become more interesting.

In the realist tradition, sovereignty is seen as a characteristic of the state qua state, and states are sovereign in two ways: they are both externally and internally sovereign. States are externally sovereign because there is no higher authority in the international order. Some might point to the UN as a higher authority, and it might be in theory, but in practice, the UN has been relatively ineffective, in spite of the many peacekeeping operations and sanctions that it has participated in or enacted.[15] The UN is currently impotent without the voluntary enforcement by individual, more powerful states. A state is externally sovereign when it establishes the boundary between the domestic and international spheres of politics. The state is the actor in international politics because the state, "rather than a religious or economic organization, is a repository of ultimate authority within the political space that is defined territorially."[16] Walzer similarly bases his argument against intervention as a violation of the external sovereignty of states.

Alternatively, states are internally sovereign because they possess the political decision-making authority of their citizenry, representing them internationally, and they monopolize violence both within their borders and of their citizens abroad. Others, such as Walzer, take this notion further and view sovereignty as being formed from the collective rights of a state's citizens. (This may be the case, but I leave the discussion for another time.) Some might further suggest that there exists a boundary between the domestic and the international, where the domestic sphere (and now it seems the international sphere as well) can be further divided between the public and the private, both internally and externally. A state is internally sovereign when it has authority claims over a range of activities *within* its political space. It specifies the particular things in which the state claims to be the ultimate authority within its territory. These internal authority claims establish authority boundaries where the state exercises sovereign jurisdiction and enforces compliance.

There is a normative aspect to sovereignty, both internal and external. States do not merely have the authority to execute control over territory, population, and means of violence, and so on; they are expected to exercise it. For a state to be recognized as a sovereign state, it must satisfy certain minimal requirements. Orend (2013, 37–8) argues that it must be internationally recognized, it must not violate the rights of other states, and it must proactively support the rights of its own citizens. Territorial boundaries further define the limits of state authorization, and within this territory, states are accountable for their citizens, both at home and abroad.

Acceptance and practice of state sovereignty changed how states interacted with each other, and for many, changed the citizen's relationship with her state. Widespread, large use of private armies eventually gave way to nationalized, public forces. The international system post Westphalia became a collective of competing (sometimes cooperating) sovereign states. Thus, although the rise of state sovereignty did not eliminate private force, together with other political factors, a reluctance, and eventually a norm, against the hiring of private forces developed as states sought to eliminate private violence as a direct competitor to the state's authority.

Monopoly of the use of force

Violence then is closely intertwined with the conception of the state and state sovereignty. In order to execute control over a territory and its population, states must also exert some control on the means of violence. One state theorist in particular not only proposed this concept, he sought to show that it was this control of violence that defined states. For Weber (1926), one of the essential characteristics of a state is that it "successfully upholds the claim to the monopoly of the legitimate use of physical force in enforcement of its order." Therefore, sovereignty is a kind of control over territory and population, but more importantly, control over the means of internal and external violence. This control is executed through the raising, maintenance, and use of military forces.

Modern state theorists have adopted Weber and sought to further elaborate on his theory. Tilly and Ardant (1975) includes "controlling the principal means of coercion within a given territory" in their definition of the state. Giddens (1985) further defines the nation-state, in part, as having "direct control of the means of internal and external violence" within "a territory demarcated by boundaries."[17] However, the monopoly over the legitimate use of force is not merely a characteristic for a state; for Weber and the others, the monopoly of force is what defines a state.

Some might already be arguing that the word "monopoly" could not be empirically correct. Certainly the tools used for applying force are a commodity (a point discussed in greater detail in a later chapter), and alliances and other regional or international organizations deploy forces. Nevertheless, as Avant (2005, 26) points out, "[Even w] hen states are trading tools of violence among themselves or allocating violence through international organizations made up of states, the monopoly over of force remains (however imperfectly) with states." Weber (1926) anticipates this objection as well, stating that: "Specifically ... the right to use physical force is ascribed to other institutions or to individuals only to the extent to which the state permits it. The state is considered the sole source of the 'right' to use violence."

The other sticking point is the concept of "legitimate." Why is it part of Weber's formulation? What purpose does it serve? In other words, one might ask whether states' control of force is legitimate merely because states have the monopoly of force by definition. Leander (2006) recognizes this point and argues that the "legitimate" characteristic is a deeply ingrained concept in Western military thought, where, according to Clausewitz et al. (1976), "war is the continuation of politics by other means." Internally to the state, police use force to impose laws agreed on through a political process; externally, the military defends the politically defined national interests. Furthermore, there are exceptions where private actors use force to pursue private interests, rather than as Clausewitz terms "politics by other means." Perhaps this concept is not so problematic. Saying that only states can legitimately use force may seem merely a function of the definition; however, the private use of force then calls into question the legitimacy of the use of force by states.

Historically, this was the problem associated with private forces, manifested by the widespread damage caused by the mercenaries. Weak rulers could not prevent others from using force within their territories, and this long-term lack of strong governmental control during the late middle ages "explains why mercenaries were able to have such a devastating effect" and were loathed. Percy (2007b, 80) continues, "The routine destruction of the countryside in peace time was a consequence of the inability of rulers to maintain armies in peacetime as well as in war."

Looking to counter those effects of privatized forces, emerging governments turned to restrictions, both legal and practical, to assert control over the use of violence, internally and externally. Legislating the use of private force and turning to public force was to "claim the authority to decide when, where, and why to use violence in the international system" (Thomson 1994, 82). Eliminating the reliance on private force was not as simple as legislation alone. Rather, the shift towards favoring public force competed with the traditional use of private forces, and while private force did not disappear, the resulting interplay of states executing their authority by limiting the use of private force and the concurrent inability to hire forces from other states helped to create the so-called monopoly of force. By raising, training, and maintaining its own armed forces of its own citizens, as we will see in the next chapter, the state came to believe that it was better able to effectively control the projection of violence from its own territory. Private actors continued to exist and were employed with differing results, but the roles of their services and the dependence of the modern states on wholly privately constituted forces gave way over time.

Therefore, private security companies today, as a resurgence of the use of private force, may impact and compete with this conception of a state's monopoly of force at least in practice. It is not only the force itself nor even the forces by which the state executes violence that becomes problematic. Private security companies as private force challenge the public-private sphere separation; the questions are in what way and to what degree. Advocates for military privatization, however, like Carafano (2008, 142), would counter that

> Even if a bureaucratic organization, such as [a] police force or an army, produces a good or service, the individuals who work in that bureaucracy are private parties under contracts negotiated either individually or through a collective bargaining organization, such as a union. Individuals are not 'owned' by the state ...

Therefore, the public-private line is blurred in practice without an increase of problems. The concept that states should have a monopoly of violence and fear of private abuse does not follow.

Carafano (2008) makes a good point. Professional soldiers also contract with the government. The question at hand may be that even if states have a monopoly on *force*, there is not necessarily a corresponding monopoly on *forces*. That is, a state could delegate authority of force (with controls of course) to either public or private forces, perhaps retaining the monopoly of force simpliciter. We should then return to this interesting idea in Chapter 4. However, delegation of the use of force or entrusting force to free agents raises other issues. Policies and procedures exist in place in Western states surrounding the checks and balances over military forces. Can similar controls be implemented over private forces? Thus far, this has not been the case. As McIntyre and Weiss (2007, 78) argue, "Success in warfare and ultimately control of the state are dependent on a military formed of one's own citizens, loyal to the state, with an interest in national unity and in keeping with the current government in power."

Notes

1 ("4 From U.S. Killed" 2004) and ("Four civilians killed in Iraq" 2004) For a good discussion on the impact of the deaths of the four Blackwater employees, see West (2005). There have been many other contractor casualties; some headlines include "Blackwater aids military with armed support," (2004); ("Four Triple Canopy Security Professionals Killed in Iraq" 2005); and Morton (2005).

2 Erik Prince (2014) discusses this event in detail.

3 See also Elsea, Schwartz, and Nakamura (2008).

4 David McCullough (2005) provides the account of how the mariners from Marblehead, MA evacuated the continental army from Brooklyn and how they also rowed Washington's Army across the Delaware River. These men were eventually incorporated into the militia. Americans foremost think of the Hessians when recalling mercenary activity during the Revolutionary War; negative experiences fighting these mercenary forces from Hesse have influenced how many view mercenaries in general.

5 For example, see Elsea, Schwartz, and Nakamura (2008).

6 See Chesterman and Lehnardt (2007), Alexandra, Baker, and Caparini (2008), Carmola (2010), Baker (2010), and Pattison (2014).

7 I hearken to James H. Toner's (1993, 33) remark that "teaching military ethics is difficult, I contend, because the somewhat arcane and abstruse field of ethics must be interpreted in soldierly circumstances that appear to be entirely disconnected from apparent bloodless, clinical, and abstract ethical reasoning."

8 See also Walzer (2008), Scahill (2007), Silverstein and Burton-Rose (2000), and Pelton (2006).

9 See also Schumacher (2006), Singer (2003), and Elsea, Schwartz, and Nakamura (2008).

10 Also see Janowitz (1960), Millett (1977), Snider, Nagel, and Pfaff (1999), Snider and Watkins (2000).

11 For example, while James Pattison (2008) explores ad bellum problems with employing contractors, he focuses only on legitimate authority, right intention, and discrimination. While his analysis is good, I go into greater detail across the tenets of the Just War Theory.

12 Hagedorn (2014) notes that Middlebury College scholar Allison Stanger first coined the phrase "First Contractors' War."

13 Shorrock (2014) writes, "INSCOM's "global intelligence support" contract will place the contractors at the center of this fight. It was unveiled on Sept. 12 by the US Army's Intelligence and Security Command (INSCOM), one of the largest military units that collects signals intelligence for the NSA. Under its terms, 21 companies, led by Booz Allen Hamilton, BAE Systems, Lockheed Martin and Northrop Grumman, will compete over the next five years to provide "fully integrated intelligence, security and information operations" in Afghanistan and "future contingency operations" around the world. Key to the war will be battlefield intelligence, surveillance and reconnaissance, known by its military acronym ISR, that is fed to U.S. pilots, drones, special forces operators and military advisers to track and kill ISIS fighters and their allies. Signals intelligence, a key component of ISR, is analyzed by the NSA and combined with imagery and maps provided by the National Geospatial-Intelligence Agency (NGA)."

14 Verkuil (2007, 14) notes that "notions of 'new sovereignty' in the international setting are outside [a state's] ambit … are intriguing propositions of 'disaggregated' sovereignty that invite various government institutions to act together internationally without the direct involvement of the states." While certainly the concepts of new and disaggregated sovereignty are interesting, they need more room for discussion and must wait for a different forum.

15 Johnson (1999) also argues against the idea that the UN is sovereign and questions whether the UN could be an ad bellum legitimate authority. I return to this point in Chapter 6.

16 Thomson (1994, 15–16) refers to this as the constitutive dimension of sovereignty.

17 Both are quoted in Thomson (1994, 7).

2 The armed contractor phenomenon

A contemporary debate with a long history

Who the hell are these guys?

In 2005 in Iraq, over

> 60 firms employ[ed] more than 20,000 private personnel there to carry out military functions (these figures do not include the thousands more that [provided] nonmilitary reconstruction and oil services)—roughly the same number as are provided by all of the United States' coalition partners combined.
>
> (Singer 2005b)[1]

In Afghanistan in 2013, there were over 108,000. This startling number raises a fundamental question: Why are they there? The 2003 Government Accounting Office (GAO) report on military contracting lists three responses: "to gain specialized technical skills, bypass limits on military personnel that can be deployed to certain regions, and ensure that scarce resources are available for other assignments" (Avant 2004).[2] For the military commander on the ground, however, this question becomes moot. The private military companies (PMCs) are there and will continue to be there, operating in and around his or her area of operations; what is important are the potential impacts on the soldiers and mission.

"Wait," one may be saying, "what is the difference between these armed civilians and mercenaries?" Both are armed; both work for money. P. W. Singer provides a good discussion of the differences in his book, *Corporate Warriors: The Rise of the Privatized Military Industry* (2003, 44–8). Certainly, modern security firms have some similarities and can even trace their lineage back through history as mercenary operations. Even some common business terms used today such as "company" and "freelance" could be traced to military ventures for hire (24), as I will discuss shortly.[3] Nevertheless, they are distinct. Singer notes that armed contractors, unlike mercenaries, are (1) organized in corporate form, (2) "driven by business profit rather than individual profit," and (3) operating and competing on the open global market (45–6).[4]

Furthermore, these private military companies can be further divided into three categories by the types of services they provide or where they operate: Military Support, Military Consultant, and Military Provider (63).[5] The lines between these categories are often blurred, as companies and their contracts cross categories (as I will explain in Chapter 3), but here I will be focusing on the third category—so called Military Provider companies—ones who provide armed contractors in the field.

Welcome to the Green Zone!

From my deployment to Baghdad, I can recall at least eight distinct armed contractor companies working in and around the Green Zone (in Baghdad, Iraq, circa 2004–2005), and this number does not include the Kurdish *peshmerga* and other security forces hired by Iraqi and other entities. These armed contractors performed a number of tasks, from guarding the entrances to different compounds and escorting convoys to conducting personal security detail (sometimes called Protective Services Detail, Personal Security Detachment, or PSDs). What is interesting is that none of them reported to my boss, the ground-owning commander, nor his boss, nor his boss' boss. These armed contractors worked for whomever they were contracted. Some were in fact employed by our government. For example, they provided PSD to key Department of State or Defense Department personnel; others, however, were hired by private corporations for protection; Iraqi leaders hired still others. This phenomenon of armed contractors was not Iraq-exclusive. Armed contractors are working around the globe, including in Africa, in Afghanistan, where they have provided PSD for President Hamid Karzai and secure the US embassy, in Japan (Weaver 2007), and even in the United States (Sizemore 2005).

However, not even the Coalition Provisional Authority (CPA) exercised full control over these armed contractors in Iraq in 2004. Twice a month, CPA staffers held an armed contractor coordination meeting to discuss policy, deconflict issues, and discuss mutually important topics such as quick reaction forces.[6] These meetings and the building of informal relationships between the armed contractors and the military became essential to deconflict battlespace and missions. For example, one requirement was to coordinate efforts of the civilian PSD, military escorts, and military quick reaction forces from different military units.

An example where miscommunication and a lack of trust could have had tragic consequences was with the employment of snipers. My unit had employed snipers atop a prominent building to over-watch approaches into a vitally important Entry Control Point (ECP). Coincidently, a VIP convoy planned to move through the area, and its armed contractor PSD had emplaced a sniper on the same building. The soldiers and the contracted sniper had the wherewithal to coordinate their efforts, but if my soldier's engagement criteria were different (and they were) from the contracted sniper's, there obviously could be issues. The privatized sniper did not know my control measures and was not operating on our radio frequencies. Fortunately, in this example, the convoy passed without incident, and we were able to emplace some coordination and communication measures between the unit and the contractors for future operations.[7]

Perceived relative deprivation

Apart from command and control coordination with armed contractors, military leaders need to become more aware of another issue, especially when retention of good soldiers has become even more important. Deborah Avant (2004) notes in her article that, "News reports on the war in Iraq have noted the relatively high salaries of contractors—some $20,000 per month, triple or more what active-duty soldiers earn," rumored to be as much as 10 times. Although retention of soldiers while deployed has been thus far successful, the lure of tax-free bonuses and a sense of patriotism goes only so far. The armed contractor companies operating in Iraq were filled with many former military personnel. Some of the more famous ones were founded by and are still operated by them (e.g., Triple Canopy). For retired personnel, this kind of employment makes complete sense. These individuals have the requisite skills, and there are many private military employees who continue to operate from a sense of patriotism. However, when a company can offer you a similar job with such a large increase in salary, it is not surprising that many currently serving military personnel are considering making their career move earlier.

Consider the British Special Air Service (SAS). The SAS allegedly had to revert to some unusual recruiting techniques to keep its ranks full because so many of its soldiers are finding employment elsewhere in PMCs (Singer 2003, 77). Some sources even allege that in 2004 there were more former SAS in Iraq working in these security firms than in the active SAS, and reports of soldiers asking for leaves of absences to work in Iraq increased (Chatterjee 2004). Having this talent in a private security company does increase its competency and credibility among military and government planners, which in turn helps their bottom line. One might argue that this is another reason to maintain the tax-free bonuses while deployed. Regardless, the lure of money is something else the military and government need to take into account.

Why is using armed contractors problematic?

"Why did you shoot my car?"[8]

In early 2004, if coalition forces damaged an Iraqi's property, the Iraqi could file a claim with the district council. However, in the incident quoted above, coalition forces did not fire on this man's car. It was an armed contractor convoy that apparently fired on the car to keep the Iraqi away from the convoy. I raise this example not to point out that particular contractor's behavior, as both the military and the security contractors were operating under a similar standing operating procedure (SOP) for a while (which was eventually phased out by Multi-National Corps-Iraq (MNC-I)). Rather, it raises another, more important issue: the legal as well as the moral status of these armed contractors. According to the Laws of Land Warfare (or LOAC), combatants are allotted certain privileges that correspond to their responsibilities during war. In turn, non-combatants are immune from deliberate targeting but must refrain from bearing arms (United States Department of the Army 1956a). Morally, we can ask whether civilians are innocent, engaged in harm, of perhaps materially contributing to the war. For example, there have long been debates over the relativeness of non-combatant immunity for factory workers who make the bombs or planes or ball bearings.[9] Even civilians residing in London, Dresden, and Berlin were targeted.

It seems unreasonable, though, to conclude that these PMC employees are non-combatants; they openly carry and use arms. (Of course, one may argue that the other, unarmed military contractors are non-combatants; this argument is not con-clusive, however.) Yet, it also seems an untenable argument to call these armed contractors "combatants" as we would the soldiers, sailors, marines, and airmen. (I will return to this problem in detail in Chapter 5.)

The problem becomes even murkier if these armed individuals are captured. As a combatant, those in the military are to be treated according to the Geneva Con-ventions as POWs as they are morally no longer a threat. A POW has certain rights; an illegal combatant has none. At best, an illegal combatant would be classified as a criminal and will then be subject to the criminal system. There are no agreed upon, international conventions to cover these new, armed contractors; the laws governing mercenaries do not apply. Therefore, their legal status remains uncertain. Recall the opening discussion about the Blackwater employees killed in Fallujah. If alternatively those four were captured and somehow brought to an international court, what would be the outcome? Both the law and the moral status of armed contractors is unclear.

This unclear distinction is further exacerbated because enemies (and civilians alike) do not distinguish between the military in uniform and the armed contractors in civilian clothes. Often, military personnel wear civilian clothes to blend with the population; similarly, many contractors wear uniforms in the performance of their jobs. The Iraqi gentleman did not care whether a soldier or a contractor shot his car: he wanted recompense.

Can you court martial a private military contractor (Singer 2008)?[10]

Beyond the moral and legal ambiguities concerning combatant status, there seems little agreement as to whom these armed PMC employees are responsible. Certainly, there is a contractual arrangement that is financially- and employment-binding; however, when contractors are involved in crimes or misconduct, there does not seem to be an overarching legal authority to prosecute or punish. Often, legal concerns are left for the home state to sort. In Sierra Leone, for example, employees of Executive Outcomes were suspected of using disproportionate force, even committing war crimes (Singer 2003, 218). Nevertheless, there was no one to pursue the allegations let alone prosecute the offenders; there was no clear legal basis for doing so.

In Bosnia, employees of DynCorp allegedly were involved in sex trading, normally punishable under the Uniform Code of Military Justice if they were mil-itary personnel. Avant (2004) notes that most

> security company activity falls outside the purview of the 1989 U.N. Convention on Mercenaries, which governs only such egregious soldier-of-fortune activities as overthrowing a government.... For example, when personnel from the US outsourcing firm DynCorp ... were implicated in sex-trade schemes, neither the contractors nor the US government was subject to international legal action.

The long wars in Iraq and Afghanistan brought out many similar issues. Until July 2004, contractors in Iraq were subject to the laws of their home country, not CPA's nor Iraq's, which made prosecutions difficult.[11] In the infamous Abu Ghraib incident, the Army's report highlighted that most of the interpreters and some interrogators, it appears, were contracted and perhaps involved.[12]

In addition, there was another armed contractor incident in Fallujah. US Marines detained sixteen Zapata Engineering contractors who had supposedly fired upon unarmed Iraqis and the Marines themselves. After the Zapata detention, stories began to appear of potential detainee abuse; these stories further complicated the differing perceptions of the events leading up to the incident. The Naval Criminal Investigative Service conducted an investigation, and each of the Zapata detainees was given a memo barring them from working in the Al Anbar province.

Although it is clear that these incidents are isolated, and the fact that soldiers have been found guilty of similar allegations, it does not eliminate the potential conflict and disturbance that these types of incidences could have in a unit's area of operations. Marine Colonel John Toolan sums up this tension well in an interview with PBS's *Frontline*: "We have a tendency to want to be a little bit more sure about operating in an environment," he said. "Whereas I think some of the contractors are motivated by the financial remuneration and the fact that they probably want to get someplace from point A to point B quickly, their tendency [is] to have a little more risk. So yes, we're at odds. But we can work it out" (Porteus 2005).

Governments no longer control the primary means of warfare

Col. Toolan raises an important point. Even if the armed contractors in a military unit's area of operations share the same desired endstate and even patriotic motivation, they may have differing immediate interests and may have a different approach to the problem at hand. Armed contractors by their very nature are profit driven (a point I will further explore). Being focused on the bottom line may not be such a negative *raison d'être*. It is this same desire to make money and the corresponding fear in the long run of losing precious contracts that provides some control over contractor conduct. One can imagine the turmoil if a company contracted with the US government reneges on, or worse violates, this contract to such a degree that the government bans further business dealings with them; the government would be in a serious bind, and the results would be disastrous for the contracted company. Nevertheless, there have been broken contracts and conflicts of interest in the past.

Stories from the many struggles in post-colonial Africa highlight these issues. Companies hire one organization, while the government hires another. Meanwhile, the rebels seeking to overthrow the government and take over the mines hire their own. For example, in Eritrea, there were former Russian and Ukrainian pilots allegedly flying for both sides. Although they dutifully bombed their respective targets according to their contracts, they evidently refused to engage in aerial combat (Singer 2003, 158). Furthermore, Columbian drug cartels were allegedly hiring armed contractors (14). In a more innocuous example, Canadian military equipment was stuck aboard a privately owned ship returning to port because of a business dispute. Fortunately, the Canadian military was returning home and not deploying for an immediate contingency, and the business conflict was resolved (160). These examples may seem farfetched, especially when discussing private security companies that are run by former US and coalition military personnel, still loyal to their governments; yet, they do happen. Plus, what about third country nationals? Many of the security companies in Iraq and Afghanistan hired local and third country nationals. These same companies would need to guarantee the loyalty of these employees, certainly keeping them from 'sensitive' environments. For the most part, the private military companies do, but this raises another area for concern.

Finally, a last issue remains. All of us who have worked for more than one employer have, in effect, left that former job for one reason or another. What happens when the armed, contracted personnel that you counted on for an upcoming event also leave for greener pastures? Singer (2005) writes,

> [Contracted] employees, unlike soldiers, can always choose to walk off the job. Such freedom can leave the military in the lurch, as has occurred several times already in Iraq: during periods of intense violence, numerous private firms delayed, suspended, or ended their operations, placing great stress on US troops.

So far, these all appear pragmatic concerns—very real, but ones that could be corrected and prevented perhaps through stronger contracting. However, there are deeper concerns with the hiring of privatized forces. That armed contractors view their work as employment like other occupations begs the question of not only how one can put a price on this kind of work but also whether this attitude reflects mercenarism, similar to that of private forces of the past. If so, is it worth the costs of employing armed contractors? Before we attempt to answer this question, it is worth seeing how we got to this point.

A brief history of privatized force

As the War on Terror, or the long war (as it is sometimes called), moved from Iraq to Afghanistan to Syria, Libya, Yemen, and the horn of Africa and the Magreb (and back to Iraq), it appears as though some of the same privatization issues that seemed prevalent and perhaps unique to Iraq have only shifted. In a July 26, 2009 article, Pincus noted that "The US military command is considering contracting a private firm to manage security on the front lines of the war in Afghanistan, even as . . . the Pentagon intends to cut back on the use of private security contractors." Earlier in that July, the Army "published a notice soliciting information from prospective contractors who would develop a security plan for 50 or more forward operating bases and smaller command outposts across Afghanistan" (Pincus 2009a). One might ask, how did we get to this point? Why does there seem to be a proliferation of privatized military?

The concept of privatized force is not a new one; rather, the privatizing of violence has had a long and sometimes inglorious history. Furthermore, the recent phenomenon is a kind of evolution—one that can both trace its history back to ancient times and intertwines with relatively modern conception of the nation-state.

While, as I will show, the US has had a long history with the privatization of military functions, the current debate owes its resurgence to the post-Cold War humanitarian interventions. "It was a tough conundrum," Singer (2003, 6) writes,

> How could the US military find a way to provide the logistics for its forces, without calling up reserves or the National Guard, while at the same helping to deal with the humanitarian crisis that the war had provoked? The solution to this problem turned out to be quite simple: the US military would pass the work on to someone else, in this case to a Texas-based construction and engineering firm.[13]

In fact, every UN peace operation conducted since 1990 included the presence of private military companies (Avant 2005, 7). To better understand the moral dimension of armed contract phenomenon, one needs to become familiar with: (1) where this conception originates; (2) how armed contractors see themselves as distinct from mercenarism; and (3) how the past provides more than a residual adverse bias. Furthermore, as Isenberg (2009) points out, categorizing the different contractor firms is not only complex but also affects the debate on whether the use of armed contractors is morally permissible. He writes,

> It is extremely difficult to generalize about private military and security firms. As an industry, or at least, business sector, PMCs have been around for less than 20 years. And although they have attracted growing attention from analysts, scholars, governments, and the general public in the past decade, there are still no agreement on how to define them, let alone categorize them.
>
> (157)

While much of the groundwork has been done by other academics, this project is noteworthy because it draws commonalities and highlights distinctions among the different views. Moreover, although it will be clear that present-day private military companies have historical mercenary ties, the private military companies should not be judged on these alone. Thus, in this next section, I set out to establish the history of privatized force and how its modern manifestation will always maintain connections with the history of mercenarism. Next, I will unpack the conception of what is a mercenary, offering both a historical perspective and an introduction to mercenaries and international law. I agree that private military companies are not mercenaries, and I will highlight what makes them distinct from their mercenary ancestors.

Ancient history of privatized force[14]

In almost every ancient history account, soldiers for hire have been present on the battlefield (often on both sides). For example, as early as the 1294 BC Battle of Kadesh, the Egyptian army included Nubian units hired by Ramses II. Not much is known of the details of these contracts, but the use of armies for hire would shape warfighting for the next 3000 years. Popular accounts of Greek military action focus on the Spartans or Athenian citizens, called to defend the honor of the respected city states. History of course points to a more complex system of warfare, where it was common practice to hire expertise in warfighting, including "Cretan slingers, Syracusan hoplites, and Thessalian cavalry" (Singer 2003, 20).

Ancient Greek historian, Matthew Trundle (2004) discusses the Greek mercenary in his *Greek Mercenaries: From the Late Archaic to Alexander.* He writes,

> When the first Greek mercenaries appeared in Aegean cannot be known. It must have been very early in Greek history because of the endemic nature of war and ancient society . . . . The first recognizable Greek mercenaries come to light it overseas service for certain during the Archaic age (the eighth and seventh centuries BC).
>
> (4)

The Greek word "Epikouri" is sometimes used to describe these ancient warriors for hire, although it is not always clear that the Greeks intended epikouri to mean mercenaries in all occasions; often it is translated as fighting companion or foreigner (Lavelle 1989, 36). Xenophon's "Ten Thousand" were Greek mercenary hoplites hired by the Persian prince Cypress the Younger for the Persian civil war (circa 400 BC). Xenophon (Xenophon and Cawkwell 1978), in his *Anabasis,* immortalized their plight, describing how the Ten Thousand, whose contract became voided once their employed was killed, made their way back to Greece.

As Trundle (2004, 6) notes, "In the late fifth century," with the Great Peloponnesian War and the Athenian defeat in 404 BC, "the numbers of mercenaries in the land forces of the Mediterranean were on the brink of an explosion . . . The Carthaginians in turn became large-scale employers of Greek mercenaries." Moreover, Alexander the Great employed over 50,000 mercenaries in 329; the Carthaginian armies were formed of mercenary units; and, Hannibal led a mercenary army when he invaded Italy (Silverstein and Burton Rose 2000, 145–6).

Machiavelli also points to the long history of mercenarism, albeit in a darker light. He writes that

> Of ancient mercenaries, for example, there are the Carthaginians, who were oppressed by their mercenary soldiers after the first war with the Romans, although the Carthaginians had their own citizens for captains. After the death of Epaminondas, Philip of Macedon was made captain of their soldiers by the Thebans, after winning the victory he deprived them of liberty (1995, XII, 36).

Medieval use of privatized force

Machiavelli's *The Prince* highlights what he perceived as the folly of using hired forces, and although he draws on the ancient past in the above passage, Machiavelli argues that because the Italian city states have long relied on mercenaries (in the late Middle Ages) they risked being weak states, doomed to fail. Mercenaries operated throughout the European Middle Ages, flourishing for a time during the feudal system. Much of what we take for granted—in practice and in terminology—in both the military and business spheres originated during this period. Medieval soldiers not employed or tied to a particular noble were known as "free lances," referring to both their employment status and choice of weaponry. As Singer (2003, 24) notes, "sooner or later the money ran out or that phase of the war came to an end. In either case, the soldiers were left without employment. Having no homes or careers to return to, many of these soldiers formed "Companies" (derived from "con pane," designating the bread that members received)."[15]

Sarah Percy (2007b) discusses the mercenary forces and their use in France and England in the twelfth to the fourteenth century in her *Mercenaries: The History of a Norm in International Relation*. Percy looks at the late medieval period as a critical phase in the formation of a general bias and norm against employment of hired forces. What defines the medieval conception of the mercenary, however, is not what many think of them today—fighting for financial reasons. Rather, it was the mercenary's lack of a justified cause for which he fought that made mercenarism repugnant (71).[16] Mercenaries were problematic because they lacked affiliation to a cause that was just, and they also fell outside of state control. The idea that mercenaries fight for different, improper causes will continue to remain relevant and evolve in the eighteenth and nineteenth centuries. And the question of private military operating outside the control of the state remains a concern for some who oppose the hiring of PMCs today. Of course, the perception that mercenaries were criminal and their conduct was reprehensible also derives from the medieval period. Percy writes,

> In France, the period of truce during the Hundred Years War, 1360-1369, left many unemployed mercenaries on French soil. With entrepreneurial zeal, they took to ravaging the countryside in companies, demanding protection money from towns and cities. Independent mercenaries 'were the scourge of Western Europe before the emergence of standing armies in the fifteenth century.'
>
> (78)[17]

The perception that mercenaries lacked a just cause and operated with a disregard for legal conduct, Percy contends, continues to affect the debate on the use of private military contractors today.

Early modern private force

In her book, Thomson (1994) succinctly argues that nonstate violence at the end of the Middle Ages was directly intertwined with a marketization of force and the global, imperial aspirations of the emerging states. She writes,

> Rulers began authorizing nonstate violence as early as the thirteenth century, when privateering was invented. Large-scale private armies dominated Europe in the fourteenth and fifteenth centuries. Mercenary armies were the norm for

the eighteenth-century European states; naval mercenaries [were used] through the eighteenth century. Mercantile companies flourished from the sixteenth to the nineteenth century. All these practices reflected the marketization and internationalization of violence that began with the Hundred Years' War.

(21)

Many of the characteristics of today's private military companies can be traced to these kinds of private force manifestations.

Pirates vs. privateers

The emergence of Europe out of the Middle Ages brought two major versions of privatizing force into widespread use: the privateer and the charter company. Both were used to further European imperialist desires, and both were often implicated in more notorious actions, which in many ways resembled their illegal cousin's actions (e.g., piracy). These shared characteristics are often highlighted when discussing contemporary private military companies. Although privateers and chartered mercantile companies do not exist today, piracy it seems has never been eradicated, as evidenced in the recent attacks off Somali waters (see Pitney and Levin 2013). While perhaps unfairly, some might contend that the private military companies of today are more like the grandchildren of the privateers and charter companies of the past.

Kinsey (2006) writes, "As far as sea power is concerned, the privatization of violence is bound up with the practice of privateering," whose origins can be traced back at least to the eleventh century. International law defines "Privateering" as "'vessels belonging to private owners, and sailing under commission of war and empowering the person to whom it is granted to carry on all forms of hostility which is permissible at sea by the usage of war.'" Therefore, Kinsey continues, "The act of privateering was thus a wartime act by which the state authorized private individuals to attack enemy commerce, and allow them to keep a portion of the prize seized as payment." For compliance with their government, the privateer vessels are often under bond, and the cargo was subject to inspection (36–7).[18]

As Talty (2007, 35) points out, it is some seemingly minor differences that separate privateers from pirates. Privateers had letters of marque or commissions (though these could be revoked); thus, they were referred to as "licensed marauders of the sea."[19] These official commissions brought with them the legal protection of the crown, and often this allowed many of the crews of the privateers to think of themselves as patriots. Because the privateers shared their treasure with the crown, were licensed by the crown, and often carried out orders from the crown, they were considered "completely respectable" (by their fellow countrymen); even nobles signed-up. Pirates, of course, had no commission and attacked anyone. Today the stereotype of pirates as solely self-motivated by the reward of plunder and a life of lawlessness continues. Ironically, members of competing countries, including their commissioned privateers, considered their advisories pirates in spite of their king's commission.

Even in the eighteenth century, problems continued to plague states that used privateers to conduct foreign policy. The first revolved around the legal definition of what pirates (vs. privateers) actually were, and what were states' obligations in regards to pirates. The second was the near impossibility of distinguishing the privateers from the pirates. The problem of accurately defining piracy under international law was not only a problem in the eighteenth century, where the US and European governments had to contend with the Barbary Pirates, but continues to be debated today, especially in cases where the pirates do not seem to belong to any particular state. Thomson (1994) writes,

> It is not clear whether piracy is a crime under international law or, if it is, when it is outlawed. The [foundational] characteristic of the pirate is that his violent acts are not authorized by a state. Thus, if piratical acts are divorced from state authority, and therefore responsibility, they do not come under the rubric of international law, which deals only with sovereign states.
>
> (108)[20]

Moreover, as I will discuss later, defining different types of privatized force, including private security companies—much like defining piracy, continues to remain unclear in the international law arena. And, if one holds as Walzer explains that soldiers have a moral war right to kill tied to their defensive service to the state, then others who operate outside of the state are morally more like pirates.

The second problem with privateers, one that also finds relevance in contemporary debates about privatized force, concerns a matter of perception. As the colloquialism goes, "one man's terrorist is another man's freedom fighter;" determining who were privateers and who were pirates was both a difficult endeavor and rather dangerous. Thomson notes, "[Who] is to say whether the Elizabethan Sea Dogs, the Barbary Corsairs, or the Malabar Angrias were pirates or privateers?" (104). Pirates would be granted commissions and return to being pirates at the end of sanctioned hostilities. Nothing apparently changed, except that in certain circumstances the pirates had a letter of the marque. Today, some might argue that even though technical differences exist between private security contractors and the armed forces, to the enemy, the insurgent, or even the innocent bystander, using certain military weapons, conducting business using certain military doctrine, operating within a specific area of operations, and even functioning under certain foreign policy all combine to blur this distinction. So, it seems, as I argue later, this problem continues today.

Charter (or mercantile) companies

As the great states of Europe consolidated their power, they turned to another, officially sanctioned privatized entity to further their imperial goals. The charter or mercantile companies, such as the Dutch East India Company and the English East India Company, began to exercise power in the sixteenth century. Officially, these companies were chartered by their respective states to regulate trade in emerging markets around the world, and were also ostensibly designed to expand both the territory and populous of the growing European kingdoms. For almost 350 years, these charter companies oversaw not just trade in far-flung colonies but also functioned as an extension, and often competitor, to their home state's government. Although organized as private businesses, and often traded with public shares through shareholders, they were officially licensed by their home country to conduct a wide range of tasks, and these tasks often mirrored the same sovereign tasks that their home governments practiced.[21] As Kinsey (2006) notes,

> Companies also had to protect themselves from attacks by local forces, pirates, or other Europeans . . . . Under such conditions, it was easy for the state to justify granting of sovereign power on the grounds that, to establish a trading relationship with its European areas, a company needed to be able to protect themselves, their ships, and goods from attack from other hostile groups interested in their business.
>
> (38–9)

In addition, it was not merely forts and ships that the companies built; they also raised and maintained their own police forces and even their own armies. While these companies grew richer and their home states expanded their power and influence globally, relationships between the parent state and their licensed companies were not always harmonious. Driven by profit, and often having a far different perspective locally, the charter companies sometimes conducted policy that resisted or even contravened their home government's intentions. Whether it was Hudson Bay Company gunners firing on Royal Navy ships or the Dutch charter company aggressively pursuing policy against Portugal over lands in Brazil, charter companies became notorious for conducting policy, while perhaps within the law's authority, outside of state politics.

One avenue proponents of private military companies might take is claiming that they are merely the next evolution of force. Advocates of this view, however, must accept the nefarious aspects of their evolutionary forebears. Charter companies— while not businesses established for the sole use of force, nor for security operations only—are examples of privatized corporate entities with a license to use force. Some might argue that the power or influence private military companies have today: (1) pales in comparison and (2) would never be used to challenge the home state; however, history suggests otherwise.

The demise of privatized force

From the above examples, one might conclude that privatized force would remain a permanent fixture in international relations. As Avant (2005, 27) writes, "Stretching from the twelfth century to the Peace of Westphalia, military contractors employed forces that had been trained within feudal structures (and were frequently licensed by feudal lords) and then contracted with whomever could pay it." As privateers competed on the oceans for their respective governments, and charter companies expanded both their own wealth and trade as well as their home country's territory, changes on the ground in Europe, in particular in how armies were organized and funded, reflected changing attitude towards privatized force.

That the reliance upon privatized forces diminished is not in doubt; however, there are a number of theories concerning why it occurred. Some argue that the formation of the nation-state caused a shift from private to public forces. Others, such as Thomson (1994), argue that the international practice of neutrality forced governments to control the means of violence; therefore, privatized force was simply no longer feasible. Finally, others suggest that the shift from privatized force was an inevitable effect of states consolidating sovereignty both internally and externally.

Each of these theories offers insight into not only why privatized forces slowly disappeared but they also delve into some of the questions that form the debate over private military companies today. Investigating the history behind this change from using privatized force reveals three themes: (1) privatized forces gave way to standing armies; (2) a norm developed against using privatized force grew out of a bias formed by what was perceived as unacceptable motivation; and, (3) the use of privatized force impacts (and may actually conflict with) state sovereignty. All three themes are highlighted in the following examples and can also be extracted from the American experience with privatized force.

Italian city state

The commercialization of war began in Northern Italy as early as the eleventh century. As Machiavelli would later discuss, military power was truly marketized, with each Italian city state hiring military and naval forces. Kinsey (2006) writes that this trend of widespread use of privatized force was enabled by the "ease of transportation and communication experience by city states ... This allowed the ready importation of skills from adjacent, more developed areas, including the Ottoman Empire and the wider Muslim world" (34). By importing these skills, the city-states could expand their commercial interests, and with the influx of wealth, the city-states could hire soldiers to fight their wars. "Thus," Kinsey continues, "there occurred a shift in the social balance in Northern Italy that favored merchant-capitalists over the knights' relationship to feudal communities" (35). Warfare became dominated by hired armies, such as the Swiss pikemen (whom I will discuss in a moment), and the Italian hired mercenaries, the condottieri, formed the military formations of the Italian city states.

Machiavelli was not the only political theorist writing about his concerns over the use of the condottieri. Swiss religious leader Ulrich Zwingli, for example, argued that "using mercenaries would undermine civic spirit, which ought to be devoted to the preservation of the city itself[,]" and as Percy (2007b, 75) notes, "[he] condemned the *condottieri* because both classical tradition and current preoccupation with active civic participation ran counter to the practice of employing mercenaries." The first problem with the condottieri is that their very existence negates the citizens' responsibility to support their state.

The other problem was more pragmatic: the administrators of the city states began to recognize that there were several competing condottieri, and they often fought amongst themselves, preyed upon the local population, or even turned on their employers. To counter these issues, the Italian city-states tried different methods of control to rein them in. First, they generated long-term contracts. Second, they established systems of muster and review, standardizing personnel (a change that will become more important). Third, they divided contracts between different groups of condottieri. While these changes helped provide a manner of control over the con-dottieri, they did not answer Zwingli's or Machiavelli's foremost objections—that a citizen has a duty to state, and the presence of contracted condottieri precluded her from doing her duty.

The Thirty Years War

Like the Italian city states, many of Europe's rulers began to recognize the benefits of using professional, standing forces—not just individual soldiers but whole units, equipped and trained for war. The first of these were commercialized, formed from the companies of out-of-work soldiery. Much like Xenophon's Ten Thousand, these units became larger and more sophisticated, and the better one's reputation, demand for their services increased. The Swiss pikemen grew to have such a reputation. Swiss mercenaries came into service in the thirteenth century and served the French crown until 1793 (Percy 2007b, 73).[22]

One could not deny the increased military effectiveness of standing, mercenary units, such as the Swiss. Nevertheless, the Thirty Years War (1618–1648) not only involved most of Europe, but again changed how armies were organized and how they fought. In addition, the large swaths of destruction—much of which was caused by mercenary units—did little to contradict the impression that forces for hire were motivated only by profit and bent on destruction. Standing military units of the period were not more effective merely because of their specialization in war, they also had the advantage that comes from unit cohesion. Kinsey (2006, 41) writes, "With the creation of a single identity within the military, soldiers now saw their units, from divisions down to platoons, as a surrogate family."

A different, privatized version of these units were the German Enterprisers. Active in the 1600s, the enterprisers raised, trained, maintained, and employed whole armies, and offered them to the highest bidder. One famous enterpriser in particular, Wal-lenstein, was on the one hand very successful during the Thirty Years War; yet on the other hand, his demanding that his colonels swear allegiance only to him raised the ire and concern of the Catholic leader Ferdinand II, who thought Wallenstein might switch sides and so had him killed. Perhaps ironically, Wallenstein was assassinated by Anglo-Irish mercenaries hired by Ferdinand, as Wallenstein was, it seems, attempting to contact his former Swedish mercenary adversaries for employment (See Percy 2007b and Thomson 1994).

Wallenstein did, however, help innovate a new concept in warfare—taxation to pay for the military. Much of the destruction of the Thirty Years War was due to the large armies pillaging and living off the land. Like the privateers on the waters, marauding armies augmented their meager incomes through ransom and extortion, which were widely accepted as unfortunate yet necessary byproducts and means of fielding large armies. Raising taxes to support the military was a new concept—a concept that seemed to support one of the new sovereign tasks relegated to the nation-state that emerged at the conclusion of the Thirty Years War. Avant (2005) writes, "[T]he development of a notion of service was based on payment, but payment ... [through] taxation by the state and regular pay for the military rather than plunder. [Taxation] reinforced the idea of professional military service and tied the fate of soldiers to the fate of the political entities ..." (250). With an alternative source of raising funds for the military, and with concepts such as unit cohesion and identity came a growing need for standardization. Units were contracted by, paid by, and fought for monarchies, "all of which needed the bureaucrat to organize" and standardize; put simply, standardization of military units made financial sense. Kinsey (2006) observes, "Thus we see, during the period 1560–1660, the amalgamation of mercenary companies into standing armies" (41).

One the largest changes to the use of privatized force came with the Treaty of Westphalia, at the end of the Thirty Years War, and with the Treaty of Westphalia came the birth of the modern nation-state. The citizenry of the nation-state were subject to that state's laws, and their loyalty was to that state. States, not religious groups, were the ones to exercise sovereign power—specifically sovereign power within a state's borders. Theorists, such as Walzer (2006a) and Orend (2013), point to the rise of (and primacy of) complementary state rights of territorial integrity and political sovereignty.[23] Raising and maintaining forces still occurred, and often a state's army included large numbers of foreigners, but states, not individuals nor cities nor free companies, took on these responsibilities.

The French Revolution and the Prussian defeat at Jena

Two events in Europe highlight the continuing shift away from large armies for hire towards the use of standing, citizen-based armies: the French Revolution and the Battle of Jenna. While tactics improved with the introduction of linear formations in battle, which allowed massed firepower with large, relatively mobile forces in the 1700s, it was the French Revolution, in particular the French adoption of levee en masse, that created massive armies, formed of a state's citizens, that together with the new tactics allowed states' reliance on commercial force to wane.

The 1793 levee en masse was more than a national draft; of course all males of fighting age were required to join the French forces, but so was the rest of the citizenry of France. The Convention of August 23, 1793, which initiated the levee en masse, read,

> The young men shall fight; the married men shall forge arms and transport provisions; the women shall make tents and clothes and shall serve in the hospitals; the children shall turn old lint into linen; the old men shall betake

themselves to the public squares in order to arouse the courage of the warriors and preach hatred of kings and the unity of the Republic.[24]

For the first time on such a large scale, the concept of a citizen's duty to protect the state was institutionalized. Through the standardization of uniforms and equipment, the introduction of levee en masse greatly increasing the size of fielded armies, and the evolution of tactics that capitalized on larger armies, the French army became the model army and a feared force on the continent of Europe. Other European armies of the time were nearly all mercenary armies. However, even for these other states,

> Mercenary armies gradually became standing armies as the financial cost of implementing the changes introduced by the military revolution made the practice of disbanding and paying off mercenary armies at the end of each campaign season, and then re-enlisting for the new campaign season, extremely costly.
> (Kinsey 2006, 36)[25]

That the French army reflected more than a mere evolution in the composition of force from the privatized sphere to the public would be demonstrated on the battlefield at Jena.

In 1806, the standing mercenary army of Prussia met French army at Jena and was soundly defeated (Citino 2005). Following their defeat, the Prussians (correctly or not) attributed France's victory to the need for a citizen-based force. Prussia rapidly re-created its army into a citizen army and was quickly able to return to flight and helped defeat the French forces. With the two largest armies on the continent (France and Prussia) demonstrating the effectiveness of large citizen armies, the citizen army became an international model. While certainly large armies, even ones formed from a state's own citizens, were risky to their rulers, their effectiveness on the battlefield made them necessary for the state's security (Palmer 1986).

Thus, the French Revolution led to a relevant revolution in military affairs (RMA) in Europe (Kaldor 2006, 152). For the first time since before the Middle Ages, states raised militaries and fought wars using their own citizens almost exclusively. Armies comprised of foreigners and mercenaries all but disappeared from the armies of Europe. The citizen army proved its worth at Jena; perhaps more telling was the almost sole use of citizen armies during the nationalist wars of the nineteenth century. This trend continued through the world wars of the twentieth century, and in spite of technological advances, continues to be the norm, and private force continued to operate but only peripherally.

Privatized force in the United States[26]

The changes that occurred in European states concerning the structure of the military and the evolution of military tactics also occurred in their colonies worldwide, and in particular in the United States. Therefore, a look into the background and a brief history of the use of privatized force in the United States is important to understand both the American reliance on hired military force as well as its long-standing reluctance to use it.

Early American privatized force

As many proponents of privatized security companies are quick to point out, the United States has always relied on some form of privatization to support its military. Even so, it was the US colonists' battles with British hired mercenary forces that form much of the lingering bias against private force. During the American Revolution, the British government was involved in other wars closer to home and was unable to send a large number of its own troops to keep the rebellious colonies in line. The English king, George III, contracted with the Landgrave of Hesse-Kassel for 12,000 troops for duty in the American colonies. Similar contracts were made with the Duke of Brunswick and other German states for another 5,000. These contracts paid an annual subsidy for service, and extra payment was to be paid for those killed or wounded.[27]

It should be pointed out that Britain was not unanimous in its decision to use Hessians in the American colonies; there were concerns ranging from the negative perception that the empire should rely on someone else to police itself to concerns that this was an internal issue, one that required an internal solution. (I will discuss some of these issues later.) Of course, because the British had turned towards the option of hiring privatized forces, the American leadership was placed in somewhat of a dilemma. The American colonies had no standing forces of their own. However, hiring of foreign mercenaries was contrary to one of the original grievances leveled against George III in the Declaration of Independence, and hiring foreign soldiers seemed to run directly counter to the republican ideals that began to shape the emerging American identity.

The American Declaration of Independence specifically mentions the king's hiring of foreign soldiers, and the colonial leadership used this hiring as one of the pretexts for declaring its independence from England. The Declaration states, "[George III] is, at this time, transporting large armies of foreign Mercenaries, to complete the works of death, this elation and tyranny, already begun with circumstances of cruelty and perfidy scarcely paralleled in the most barbarous ages and totally unworthy of the Head of a civilized nation." Furthermore, the language in the document captures the long-held bias, from Machiavelli's writings to experiences during the Thirty Years War, that mercenary force in and of itself manifested death and destruction.

This does not mean that the American leadership never considered hiring forces or hiring of military experienced leaders to help in its revolution. Logistical support was certainly hired, and they were thrilled to get help from anyone. One foreign hire was Thaddeus Kosciusko, a Polish officer who served as a general in the American forces. As an engineer, he helped fortify the revolutionary capitol in Philadelphia, and he commanded the military engineering works at West Point. (His monument stands at the United States Military Academy at West Point, NY, and other statues to him exist in Detroit, Boston, and Washington DC.) Other notables include the Marquis de La Fayette (Lafayette), Kazimierz Pułask (known in the US military as "the father of American cavalry"), and Friedrich Wilhelm von Steuben, who is remembered for teaching the American forces military drill and discipline. In addition, the American leadership relied extensively on privateering for its naval forces until a permanent Navy could be established.

Nevertheless, despite the hiring of support forces, as well as employing foreign military expertise, the colonists—and later the United States—relied on the use of local militia and later the citizen Continental Army to defeat the British and earn its independence. As Percy (2007b) highlights, "the newly declared state identified itself as a republic made virtuous by the use of a citizen army, and doubly virtuous because it fought against mercenaries" (25). The American public during the war, and later, after it won its independence and became a state, thought that the matter of considering whether to choose hired forces was closed. The American identity was tied to the popularized image of the Minuteman at Concorde, and the use of privatized force would have contradicted the American identity.

Post-Revolution privatization

The American experiment with privatized forces did not, however, end with the American Revolution. In the early years of the United States, individuals, some businessmen, some employees of the government, launched military expeditions from the United States territory and were known historically as filibusters. Some of these filibusters seem to be motivated by chance for increased personal gain, either financially politically or both, and in several cases seem to operate in an albeit unofficial capacity to expand the territory of the United States. Some filibusters even attempted to establish independent states with them as the head of state (Thomson 1994 and Ryan 2008).

United States also faced a more pressing problem during the French Revolution. The United States, although allied with the French during the Revolution, maintained its neutrality during the fighting between the French and English. Nevertheless, American merchant ships were being raided, and their sailors impressed into service (altercations that eventually lead to the War of 1812), and European governments, such as France, were actively recruiting in the United States. As I discussed earlier, part of the American identity was tied to the republican ideals of a citizen's duty to the state as well as the state's responsibility to its citizens. Thomas Jefferson captures this argument succinctly; he wrote, "the granting of military commissions, within the United States, by any other authority other than their own, [is] an infringement on their sovereignty, and particularly so when granted to their own citizens, to lead them to commit acts contrary to the duties they owed their own country" (quoted in Thomson 1994, 77). By recruiting military on US soil, the French not only violated the sovereignty of the United States but also challenged these republican ideals. Therefore, the US passed the Neutrality Act of 1794, which prevented "citizens or inhabitants of United States from accepting commissions or enlisting in the service of a foreign state."

Yet, in spite of the apparent disapproval of the use of hired forces, the US, with its fledgling armed forces and Navy, had to turn to hired forces to achieve some foreign policy ends. One of the more famous of these occurred during the US's struggle with the Barbary corsairs at the turn of the nineteenth century. The US's early forays into the Mediterranean for trade began to set the tone for long standing interactions with the Middle East. Pirate navies based on the Barbary Coast and other places along North Africa repeatedly captured and ransomed American merchant ships. Much of the early US Naval history is entwined with this conflict. In one action to remove this threat, the United States launched a force under the command of Lt. Eaton to march to Tripoli and negotiate a settlement. "From the Halls of Montezuma, to the shores of Tripoli ..." begins the official US Marine Song, honoring the fight to Tripoli. From December 1804 to April 1805, Eaton and his force attacked across North Africa, taking the city of Drana. However, the Marines never got to Tripoli (the conflict was settled before Eaton's arrival). More relevant was Eaton's force composition. Eaton's "army" did include a force of Marines, but it was only eight Marines. The bulk of his force consisted of sixty-three European mercenaries (as Blackwater's founder Erik Prince is proud to point out),[28] and several hundred hired, armed locals (Oren 2007, 66–9).[29]

Nevertheless, the norm in the United States was to turn away from using privatized forces and rely on a citizen army and navy. Much of the early force structure debate revolves around the question of whether America should rely on militia forces or run the risks of maintaining a strong, standing military force.[30] Verkuil (2007) notes that, "The only reference in the Constitution arguably relevant to delegation to private parties is the Marque and Reprisal Clause. That clause at one time contemplated using 'privateers' to act for the government" (103).[31] The dichotomy of a reliance on citizen forces with the general prejudice against use of privatized force, with the occasional practice of hiring military force or expertise, however, has continued. During the Civil War, the United States government hired the Pinkerton Detective Agency to conduct intelligence operations against the Confederate forces. In the twentieth century, groups such as the Flying Tigers, and more currently operations such as Air America in Laos, demonstrate America's periodic reliance upon hired force to try to meet its foreign policy goals.

Modern mercenaries

Although the use of privatized force never completely disappeared in the twentieth century, the 1960s saw an increase in the publicized use of mercenary forces in places like Africa. Angola, Biafra, Congo, Rhodesia, and Seychelles were all countries that have become associated with the widespread use of mercenaries. Men like Mike Hoare and Bob Denard became famous for their mercenary activities in these countries (Percy 2007b, 185–9; Silverstein and Burton-Rose, 146–51). Collectively known as "*Les Affreux,*" these mercenaries became popularized in mainstream media in movies such as the 1978 movie *Wild Geese*, Frederick Forsyth's novel, *The Dogs of War*, and even in a Warren Zevon song, "Roland the Headless Thompson Gunner."

Decolonialization and the civil wars in Africa created several opportunities for these soldiers of fortune to earn their keep. However, the same negative bias against using mercenaries and private force prevalent in Europe and United States began to grow in Africa. Many African leaders blame the use of hired forces for much of the devastation of the civil wars, and some grew to view the use of mercenaries, in particular white European personnel, as imperialistically motivated and another example of outside, unwanted influence in African affairs. MacIntyre (1984) writes, "In Africa, in particular, being a mercenary meant being an instrument of colonial resistance to self-determination, of support for secessionist movements, and general destabilization" (67). In response to the use of mercenaries in places like Africa, and probably acting on the perceived effect of mercenaries as a continuation of colonial and an imperial policy, the international community published the Additional Protocol I to Article 47 of the Geneva Convention (1977), which attempted to define and criminalize mercenaries. The UN later established the International Convention against the Recruitment, Use, Financing and Training of Mercenaries in December 1989 in an attempt to institutionalize and promulgate both international and national conventions against the use of mercenaries. Nevertheless, today these international laws are understood to be ineffective in regulating the use of privatized force. I discuss why in greater detail in the next chapter.

Notes

1 CORPWATCH's "State Department List of Security Companies Doing Business in Iraq," from February 15, 2005, listed 30 private military companies in Iraq; curiously several groups were missing, including Blackwater Security.
2 I am not going to discuss whether the armed contractors are meeting the expectations outlined by the GAO; that discussion is beyond the scope of this paper.
3 Singer devotes an entire chapter to the historical rise of private military companies (he calls them "private military firms") on 19–39 (2003).
4 The distinction between "mercenary" and "armed contractor" is essential to private military proponents. I will spend some time analyzing these terms in Chapter 3 of the project. These private military firms are also known as Private Military Companies (PMCs) in other literature. Cf. Carmola (2010).
5 Singer (2003) provides an explanation of these three groupings; see pp. 92–100. He uses the examples of three PMCs for illustration: Executive Outcomes, MPRI, and Brown and Root. See also "Outsourcing," where Singer (2005b) writes, "The industry is divided into three basic sectors: military provider firms (also known as "private security firms"), which offer tactical military assistance, including actual combat services, to clients; military consulting firms, which employ retired officers to provide strategic advice and military training; and military support firms, which provide logistics, intelligence, and mainten-ance services to armed forces, allowing the latter's soldiers to concentrate on combat and reducing their government's need to recruit more troops or call up more reserves."
6 The author participated in many of these private security company working group meetings.
7 Author's conversation with contracted sniper's supervisors, May 2004.
8 Asked to author by Iraqi man whose car was shot by automatic fire from a PSD escort. June 2004.
9 There are too many references on this debate to list. Some excellent sources on this discussion include Elizabeth Anscombe's "War and Murder" (1970) and Michael Walzer's *Just and Unjust Wars* (1977, 2000, 2006a).

10 Although those PMCs who are hired by the Defense Department now fall under the Uniform Code of Military Justice, this issue is far from clear. The 2008 Defense Appropriations Bill (DAP) does extend military justice to DOD contractors but only to a limited degree. The 2008 bill did not specify that contractors may now be court martialed, which is a form of legal process tyat is reserved for military personnel. Rather, it clarified that DOD civilians and contractors are subject to regulatory guidance under provisions of Uniform Code of Military Justice (UCMJ). However, they would be tried under the Military Extraterritorial Jurisdiction Act (MEJA): §18 USC.3261: "Criminal offenses committed by certain members of the Armed Forces and by persons employed by or accompanying the Armed Forces outside the United States." The 2008 DAP also only covers DOD contractors and does not necessarily extent to contractors employed by other government organizations or by private companies.

11 Avant (2004) writes, "Even US legislation created to address this issue (the Military Extraterritorial Jurisdiction Act of 2000) lacks specifics and entrusts the US Secretary of Defense with initiating prosecutions. Countries that opposed the war may have a particularly hard time prosecuting contractors for crimes committed in Iraq. That is especially true of countries such as South Africa that claim contractors from their country are exporting services without the government's permission." Singer (2005) writes in "Outsourcing" that, "not one private military contractor has been prosecuted or punished for a crime in Iraq (unlike the dozens of US soldiers who have), despite the fact that more than 20,000 contractors have now spent almost two years there. Either every one of them happens to be a model citizen, or there are serious shortcomings in the legal system that governs them."

12 For a good discussion of the impact of the ability to properly control the behavior of armed contractors, see Singer (2005). He notes, "According to reports, all of the translators and up to half of the interrogators involved were private contractors working for two firms, Titan and CACI. The US Army found that contractors were involved in 36 percent of the proven incidents and identified 6 employees as individually culpable. More than a year after the incidents, however, not one of these individuals has been indicted, prosecuted, or punished, even though the US Army has found the time to try the enlisted soldiers involved. Nor, has there been any attempt to assess corporate responsibility for the misdeeds."

13 He continues, "Instead of having to call up roughly 9,000 reservists, Brown and Root Services was hired. Not only would the firm construct a series of temporary facilities that would house and protect hundreds of thousands of Kosovars, but it would also run the supply system for US forces in the region, feeding the troops, constructing their base camps, and maintaining their vehicles and weapons systems" (6).

14 For other concise histories of private military companies see Singer (2003, Chapter 2) and Carafano (2008, Chapter 1).

15 Sir John Harkwood's White Company, an English free company, was one of the more famous examples.

16 As I will discuss later, mercenaries have been derided because of fighting for money, being foreign, or fighting without a just cause. Yet, as Percy (2007b) notes, "it is hard to see that nationality was the defining characteristic of a mercenary in the medieval period, even if it was important. A financial motivation is similarly unable to mark what made a mercenary. It 'was not the taking a pay that was reprehensible. Payment of fees and wages was central to the militant organization of the period, and was not a matter for criticism.'" Furthermore, it seems clearly possible for a state with a just cause to resort to war hiring mercenary armies to achieve its just cause. Even so, the problem concerning state control over mercenary armies remains.

17 Percy also writes, "As much as independent, self-motivated mercenaries lying outside the control of their masters were a threat to a social order, but also constituted a practical threat to land and lives. In twelfth century France, *routiers* and *cotereaux* pillaged the

countryside, particularly in the Limousin and the Auyergne, and were known for attacking churches . . . " (79–80).

18 Kinsey (2006) writes, "In 1243, Henry III issued the first privateer commissions, which allow for the king to receive half the proceeds of any prize taken by vessels authorized under the commission." (37) See also Thomson (1994, 22–6).

19 Talty (2007) highlights the differences between pirates and privateers (35–6 and 70–1).

20 For an in-depth discussion of piracy and privateering, also see Thomson (1994, 44–55 and 106–18) and Kinsey (2006, 36–8).

21 Singer (2003, 34–7) compares the Dutch East India Company and the English East India Company. See also Avant (2005, 27–8), Kinsey (2006, 38–40), and Thomson (1994, 32–41, 59–67, and 97–105).

22 Another mercenary unit was the Landknecht, formed in Germany, which competed and fought against the Swiss pikemen.

23 Neither of them uses the theory of state rights to argue for or against the use of privatized military, but both view these state rights as integral to modern notions of just warfare.

24 *Les jeunes gens iront au combat; les hommes mariés forgeront les armes et transporteront les subsistances; les femmes feront des tentes et serviront dans les hôpitaux; les enfants mettront le vieux linge en charpie; les vieillards se feront porter sur les places publiques pour exciter le courage des guerriers, prêcher la haine des rois et l'unité de la République.* For a discussion on how this decree affected the modern idea of civilians on the battlefield, see Robert Schütte (2014, 72).

25 Ironically, this same reasoning—that there are more cost effective and more efficient military force solutions—is used today to argue for outsourcing certain military functions. It seems surprising that in the nineteenth century mercenary army costs increased to the point where standing, public forces became more cost effective.

26 Other states, namely Great Britain, have had a colorful history of using privatized force. These are discussed in greater detail elsewhere, so here will focus on the US experience with privatized military.

27 Cf. Avant (2005, 22, FN 56); Silverstein and Burton-Rose (2000, 146), and Singer (2003, 33).

28 Prince mentioned Eaton's force as an example of the US's reliance on private force at a panel discussion at the 2009 McCain Conference, April 2009, at the US Naval Academy, Annapolis, MD.

29 Thomson (1994) explains the composition in greater detail. Eaton's force, she writes, consisted of "'a company of 38 Greeks . . . about 400 [Arabs] . . . a few British subjects, two or three Germans, Italians, Spaniards, and various kinds of Levantines.' 10 Americans are also involved: Eaton, a US Navy midshipmen, and eight Marines, including a lieutenant and a sergeant" (197).

30 The debate between supporter of a militia and a standing army is well captured by Allan Reed Millett and Peter Maslowski in *For the Common Defense* (1984).

31 Verkuil (2007) continues, "It has important limitations. The Constitution requires congressional approval for implementation (Article I, Section 8) and forbids states from issuing them (Article I, section 10)." Also, privateers have not been considered as an option since Andrew Jackson; however, the clause was referred to during the Iran-Contra investigation.

3 Mercenaries, soldiers, and armed contractors

An explication

> Nowadays, people tend to label anyone who carries a gun while not a member of a regular military establishment a mercenary.
>
> (Isenberg 2009, 5)

Callsign "Ass Monkey"

We were introduced to Blackwater's light aviation unit in Iraq in 2004 because they were one of the private security groups operating in our area. Consisting of seasoned pilots, many of them former Task Force 160 special operations pilots, Blackwater's aviation, who were known by the callsign Ass Monkey on the radio, flew H-6 Littlebird helicopters providing command and control for the Blackwater protective details. They provided overhead communications, reconnaissance, and also weapons—not like an attack helicopter—but they had former special operators who were trained to shoot from a moving helicopter to provide additional security should one of their convoys be attacked. They had a reputation for buzzing around the Green Zone, even allegedly flying under the crossed Sabers monument. They seemed a relaxed group on the ground, even chill. But, I knew them in a different, professional light. They were very competent pilots, and they helped us out on more than one occasion.

One day there was a vehicle borne improvised explosive device (VBIED) incident just outside of our operating area, right across the river. We put our reaction forces on alert and prepared to move to the site. By now we had become well acquainted with VBIEDs (overall we would respond to over 20), and we had trained and developed tactics to secure the VBIED site from further attack, for exploitation, and to allow Iraqi first responders to evacuate casualties. Our methods included working with US government organizations as well as other private military companies who might be in the area or who provided heavy equipment and barrier material for clean up. We were centrally located and could provide a spectrum of response forces depending on the situation.

Rarely does one unit cross another's boundary, but in a case like this, where the other unit was much further away, it made sense that we would get the call to assist, and we did. Each incident is unique, and this one involved one private military group who appeared to be the target. Additionally, unlike in other VBIED incidents, as we were working our way to the scene, a crowd was forming around the site—a crowd that might make for a lucrative enemy target. We needed air support, but the US air patrol was engaged elsewhere. Without us asking, Ass Monkey arrived on scene and reconned the site for us. They provided the needed aerial perspective of the situation, and they stayed on scene working with us until other forces could arrive.

Working with Ass Monkey[1] that day was a great example of how private security companies and the military can work together. But, why did they come to our aid? They were not contracted to provide support to us. They were not being paid extra for this mission. Nor was it just a matter of expertise, professionalism, or even desire for the same endstate—the successful completion of the mission. They flew over us because they saw it as their duty—the right thing to do when your comrades are in the thick of it. It was a sense of common purpose and shared experience. Yet, if Isenberg is correct, these aviators were mercenaries. This label does not seem accurate but it also seems disingenuous to say that they were mercenaries that day. The men who flew in support of us did not seem to be like the mercenaries of old.

Proponents of privatized forces argue that part of the lack of their acceptance is that these new corporations are often labeled "mercenary," and—they continue— the word "mercenary" carries negative connotations, reinforced by history's negative bias against mercenaries and supported by popular media's glamorizing of the more nefarious aspects of mercenary conduct. Some might argue against private military companies based on the belief that profit motivation is somehow evil; certainly, one might draw comparison between the market motivations of private military companies and mercenaries. Taken a step further, detractors of privatized military might suggest that, like mercenaries, private military companies also operate with a lack of morality. Singer (2003) notes, "In the public imagination, they are men depicted in such films as 'The Wild Geese' or 'The Dogs Of War'–freelance soldiers of no fixed abode, who, for large amounts of money, fight for dubious causes ... In the general psyche, to be 'mercenary' is to be inherently ruthless and disloyal" (40). While these views are simplistic, overly dramatic, and truly not helpful in the debate over privatized military employment, one cannot deny that modern private military companies do have historical ties to the mercenaries of the past, and both share characteristics in common.

Identifying these characteristics would be informative to the larger debate, but there are little agreed-upon definitions—both legal and academic—to draw out accurate comparisons. Some commentators (such as Avant 2005, 23) have gone as far as suspending their analysis of terms such as mercenary. However, I think some effort exploring these terms is necessary for participating in the discussions of the ethical dimensions of privatized force. In other words, one cannot speak to how private military companies are different from mercenaries without first attempting to define them. Furthermore, defining what we mean by mercenary of private contractor is not merely a semantic game; as an analytic project, identifying the necessary or sufficient conditions that differentiate one from the other is critical for the debate over private force. Therefore, I will begin with the question of "What is a Mercenary?"

The m-word

One astute observer offered that, "The truth is that defining a mercenary is a bit like defining pornography; it is frequently in the eye and mind of the beholder" (Isenberg 2009, 6). However humorous, his comment, nevertheless, speaks to the problem of explicating a definition of "mercenary." Apart from historical accounts, highlighting stories of mercenaries and their activities, as well as popular culture's portrayal of mercenaries, there are basically three avenues to turn to when searching for a definition of "mercenary:" (1) dictionaries, (2) international law, and (3) other academics interpretations. And, these definitions are in flux.

A quick review of dictionary definitions of mercenary reveals that they are as plentiful as they are somewhat uninformative. *Webster* defines mercenary as "one that serves merely for wages; especially a soldier hired into foreign service." Or, as an adjective, *Webster* notes that to be mercenary means "serving merely for pay or sordid advantage" or "hired for service in the army of a foreign country." The *Oxford English Dictionary*, on the other hand, offers at first glance a more complete version. It defines mercenary as "1. A person who works merely for money or other material reward; a hireling" or "a person whose actions are motivated primarily by personal gain, often at the expense of ethics." Further, it states "A person who receives payment for his or her services. Chiefly and now only: [specifically] a soldier paid to serve in a foreign army or other military organization." As an adjective, mercenary describes

1   "a person, organization, etc.: working or acting merely for money or other material reward; motivated by self-interest,"
2   "of conduct, a course of action, etc., or its motivation: characterized by self-interest or the pursuit of personal gain; prompted by the desire for money or other material reward; undertaken only for personal gain," or,
3   "being hired, serving for wages, designating a soldier paid to serve in a foreign army or other military organization; (of an army) composed of such soldiers."

While these dictionary definitions seem to capture some intuitions about mercenaries, such as motivation, some important characteristics seem to be lacking. First, the notion of fighting for personal gain or to be purely profit motivated seems too broad of a characterization; in a way, any person could act out of such motivation, soldiers included. Second, specifying that mercenaries fight in foreign armies limits the scope of their employment. As noted in the last chapter, soldiers for hire were often used within a regime, and some foreign forces are part of a state's military (e.g., Gurkhas) and are usually considered exempt from the mercenary status. Today it is conceivable that a state's military might be wholly constituted by a Private Military Company (PMC). These definitional characteristics do not fully capture the idea of buying and selling military services, especially in a larger, corporate nature.

And, although the *Oxford* dictionary does make note of the mercenary's seeming lack of ethical conduct, this characteristic seems to not only reflect a pejorative bias but is also not empirically based. Nevertheless, for now, one can cull from these definitions that a mercenary is one who (1) acts from financial motivation, and (2) fights for a foreign army. Thus,

MERC: S is a mercenary if and only if she

1   acts from financial motivation, and
2   fights for a foreign army.[2]

Alternatively, one might suggest that the only definition of mercenary that counts is the legal one. The actual definition in international law is set out in Additional Protocol I to Article 47 of the Geneva Convention (1977). A "mercenary" is defined as any person who:

1   is specially recruited locally or abroad in order to fight in an armed conflict;
2   does, in fact, take a direct part in the hostilities;
3   is motivated to take part in the hostilities essentially by the desire for private gain and, in fact, is promised, by or on behalf of a Party to the conflict, material compensation substantially in excess of that promised or paid to combatants of similar ranks and functions in the armed forces of that Party;
4   is neither a national of a Party to the conflict nor a resident of territory controlled by a Party to the conflict;
5   is not a member of the armed forces of a Party to the conflict; and
6   has not been sent by a State which is not a Party to the conflict on official duty as a member of its armed forces.[3]

Furthermore, under Article 47 of Protocol I (Additional to the Geneva Conventions of 12 August 1949, and relating to the Protection of Victims of International Armed Conflicts) the first sentence states: "A mercenary shall not have the right to be a combatant or a prisoner of war." This seems a striking statement, and I will address it and its implications in greater detail in Chapter 5. In spite of the historical condemnation of mercenarism, its eventual decline of widespread usage by states, and the legal formulations against it in international law, mercenaries were used throughout the 1960s and 1970s. In addition, as I will point out below, privatized military was used throughout the late-twentieth century. Responding to current international law, a noted privatized military supporter was heard to say that if these individuals or companies get caught under the anti-mercenary laws, then they deserve what they get (Bicanic and Bourque 2006).[4]

While international law may have been relatively impotent in eliminating mercenarism overall, additional international law has been written since Article 47 to solidify the current convention. On December 4, 1989, the United Nations passed resolution 44/34, the "International Convention against the Recruitment, Use, Financing and Training of Mercenaries," and it entered into force on October 20, 2001 as the "UN Mercenary Convention." Article 2 makes it an offense to employ a mercenary and Article 3.1 states, "A mercenary, as defined in Article 1 of the present Convention, who participates directly in hostilities or in a concerted act of violence, as the case may be, commits an offense for the purposes of the Convention" (United Nations 1989). In a sense, then, being a mercenary and acting in such a capacity are now international offenses.

From the Geneva Convention and international law, we can now expand our definition of "mercenary" to include: (3) is neither a national of nor a member of a Party to the conflict, (4) takes a direct part in the hostilities, and (5) the compensation is "substantially greater" than comparable military members of a Party to the conflict. Thus MERC can be modified to MERC':

MERC': S is a mercenary if and only if she

1   acts from financial motivation,
2   fights for a foreign army,
3   is neither a national of nor a member of a Party to the conflict,
4   takes a direct part in the hostilities, and
5   gains compensation is "substantially greater" than comparable military members of a Party to the conflict.

This definition, however, now seems rather unwieldy. As Avant (2005) succinctly notes, "The fact that what 'mercenary' refers to has changed over time is interesting for what it tells analysts about the shift in what are considered legitimate uses of force, but makes the word less useful as an analytic term" (23). There have been several attempts by academics to refine the definition of mercenaries. Percy (2007b) begins her work on mercenaries, noting that a "mercenary can be defined as an individual soldier who fights for a state other than his own, or for any non-state entity which he has no direct tie, in exchange for financial gain" (53). Nevertheless, Percy's formulation seems to eliminate too much; we would be back to MERC. Singer (2003), alternatively, proposes another, more fruitful formulation. He suggests that mercenaries have the following characteristics: foreign, independent, economic motivation, oblique recruitment, temporary in nature, and simplified service objective, such as, combat. He further explains these characterizations:

Foreign: a mercenary is not a citizen or resident of the State in which he or she is fighting.

Independence: a mercenary is not integrated (for the long term) in any national force and is bound only by the contractual ties of a limited employee.

Motivation: a mercenary fights for individual short-term economic reward, not for political or religious goals.

Recruitment: mercenaries are bought in by oblique and circuitous ways to avoid legal prosecution.

Organization: mercenary units are temporary and ad hoc groupings of individual soldiers.

Services: lacking prior organization, mercenaries focus on just combat service, for single clients.

(43)

At this point, we can define "mercenary" as follows:

Mercenary: S is a mercenary if and only if she

1   is not a citizen or resident of the state in which she is fighting,
2   is not integrated in a national force,
3   fights from a financial motivation,
4   is obliquely recruited to avoid legal prosecution, and
5   is organized into a temporary, ad hoc group of individual soldiers, and this group has a simplified service objective, such as combat.

While perhaps not the ideal formulation, I suggest adapting it for this project both as a baseline for defining mercenaries, and, as I will turn to in the next section, this definition will be useful for characterizing private military companies as well.

Private military companies, not mercenaries

Private military companies vs. mercenaries

Armed with a definition of a mercenary, we can now look at private military companies, identify their characteristics, and draw a comparison. The private military industry and its advocates constantly attempt to draw a sharp distinction between their companies and mercenaries and other manifestations of privatized force (e.g., mercantile companies). One characteristic that they seem to share, however, is the motivation for profit. Whether this motivation is singular or even the primary one among others (I know many private contractors who claim that they operate out of a duty to one's country or for other motivations: the Ass Monkey pilots, for example), it is the financial comparison that seems to stick. Nevertheless, there are other differences between private military companies and mercenaries. Two important distinctions involve their (1) permanence and (2) corporate nature. Tim Spicer points to these characteristics as defining their differences.

He argues that mercenaries "are usually individuals, recruited for a specific task. They have no permanent structure, no group cohesion, no doctrine, and no vetting procedure. Their standards, both behavioral and technical, are somewhat suspect and their motives can be questionable."[5] And, alternatively, a

> private military company is a permanent structure with a large number of people on its books. It has a permanent presence, it has an office, it uses promotional

literature, it has a vetting system, it has a doctrine and it has a training capacity, internally as well as externally. It draws on a normal support that you expect from a business; it is the official military transformed into a private sector in a business guise.

(67)

It may seem strange to some to consider a temporal difference or the presence of a corporate structure as sufficient to distinguish private military companies from mercenaries, but these two characteristics bring other factors to bear.

Private military companies as corporations

Unlike history's mercenaries, private security is organized as corporate entities, which sets today's private military companies apart from mercenaries and other forms of privatized forces. First, the corporate nature preserves a kind of permanence that was missing in earlier mercenary organizations. The temporary nature of the earlier mercenary organizations was much as it was portrayed in Forsyth's *The Dogs of War*. They were based on personal networks of varying complexity. Bonds formed in prior military service and in earlier contracts formed the basis of a Rolodex-based organization. When word arose of a potential job, these networks were activated, friends called, favors called in. And, once the job was complete, the Rolodex gets updated, and all return to their other lives (Silverstein and Burton-Rose 2000, 169; Kinsey 2006, 15). Because of the size of these organizations and the temporary nature, most contracts were far removed from government-toppling coups. Rather, they performed small scale security and consulting jobs. This same small scale and temporary nature prevented these groups from earning larger and more complex contracts. Fly by night operators would give way to corporations.

Of course, permanence in the corporate world is relative. One only need look at the demise of larger companies (e.g., Lehman Brothers) to note that these large companies rise and fall; consider the money and effort that the government put into saving companies such as AIG in 2008. However, the larger companies are not established for only a single contract, even if that contract is a multibillion dollar one. For example, Brown and Root Services was awarded the logistics contract for supporting the Bosnia and Kosovo operations, but this was not a new company formed for this one contract. Brown and Root Services grew from an oil production business in Texas; and then, like many other corporations, it expanded into other areas.

Today's private military companies are more or less corporations. They usually have established and clear executive levels. They operate with a board of directors, and they often offer shareholdings. Some of these private military companies are privately owned, such as Academi (formerly Blackwater USA then Xe), while others such as Military Professional Resources Inc. (MPRI) have been purchased by publicly traded companies like L3, or they are like DynCorp, which grew out of the pre-existing military industry. Often, proponents of private military companies argue that because of their corporate nature these private military companies are accountable to shareholders, which acts as a system of checks and balances over this industry.

A second distinction between private military companies and mercenaries is that although both are financially motivated, because the private military companies are accountable to shareholders, they are driven by business profit. In addition, private military companies conduct business with larger financial holdings in corporate conglomerates to cross industry lines. This gives these new companies global reach. Today, companies such as Triple Canopy or ArmorGroup operate globally, recruit globally, can be financed globally, and may, as critics point out— be employed globally. As Kinsey (2006) argues, these private military companies retain, like other international corporations, the ability to "transform themselves from state to state, establishing parent companies, or operating subsidiary companies" (15).

Nevertheless, modern private military companies have retained a portion of the mercenary organization's temporary nature. The corporate structures of these private military companies are often hard to discern. While many of these companies are well known and their corporate officers are often in the media spotlight, a majority of the people who work for them serve limited scale contracts that are temporary in nature. The upper management of the company, the legal team, and the office managers, usually only make up a small permanent staff. The Rolodex networks of the old mercenaries have transformed into the database networks of ex-service personnel today. This feature is especially revealing when work is subcontracted or even sub, subcontracted, with many employees often residing on multiple companies' databases. Two problems arise from this database structure: first, vetting can often be difficult; second, the chain of responsibility for a given contract is often hard to trace. I will discuss both of these issues in the next chapter.

Why private military companies are not mercenaries

Some authors, such as Carafano (2008) and Isenberg (2009) (and perhaps Avant (2005)), accept the distinction between private military companies and mercenaries on face value and as uncontroversial. Other, perhaps more sensationalist authors, such as Scahill (2007), imply the private military companies are no different than mercenaries. In this section, I take a closer look at what distinguishes private military companies from mercenaries. While I do agree with Carafano and Isenberg in saying that the private military companies are not the same as mercenaries, I think some work is needed to explain why. For example, one might outright claim that the private security contractors who worked in Iraq and Afghanistan are not mercenaries.[6] To take the position that these private security contractors merely do not meet the legal definition of mercenary, as Isenberg (2009) does, is overly simplistic and does not seem to address what many intuit, that either these private military companies are no different than mercenaries or if they are different, they are no better.

One visible, perhaps trivial, way that private military companies have distanced themselves from their mercenary past is through some of the language that they use. For example, private military companies began to describe their employees as security advisers, security guards, and private security contractors, not as mercenaries. In a sense, short-term, cash based contracts are no longer suitable for private military companies. Private security, organized as a business, could yield greater profits and establish growth and future contracts. Indeed, in part due to continued attempts to distance themselves from their mercenary past, representatives of the private military industry, as part of a UN Working Group on private military companies and international law, spoke to the desire to drop the word "mercenary" altogether (UN Document a/60/263, quoted in Percy 2007a, 26).

However, if this line of reasoning is accurate and complete, then the argument concerning the differences between mercenaries and private military companies seems to reduce to one of mere semantics and trivializes the very real differences between the employees of today's private military companies and mercenary forces of the past. There is of course one major difference between the two: mercenaries historically only sold one service, participation in combat, and military companies, as I will discuss in detail next, are involved in providing a wide range of military services not just forces for combat.[7]

At this point, then, we should take note of some of the other characteristics of the private military companies. First, private military companies are organized along a corporate structure and, therefore, are permanent in nature. They are legal entities, and they often operate on the open market. They have boards of directors and shareholders. They are often linked to large corporations and other industries. Their services are traded internationally, and much of their recruitment is public. Finally, private military companies are financially motivated, but this financial motivation is driven by a desire for business profit rather than the financial self-interest of individual employees.

Therefore, based on our previous formulation of "mercenary," we can now offer a definition of "private military company":

Private Military Company (PMC): A company, C, is a PMC if and only if C

1   is organized along corporate lines,
2   is not integrated in a national force,
3   operates from a business profit motivation,
4   is a legal entity operating on the open market as its own entity or through a parent corporation,
5   is organized into an established, permanent entity, and
6   offers a range of military related services (e.g., training, advising, security).

From these two definitions, the distinctions between mercenary organizations and PMCs become more apparent. First, the corporate nature of the PMC attempts to establish it as not only a legal entity but also one that is more acceptable. Second, while the PMC is not integrated into a national force, they often operate only with direct approval, or perhaps with tacit consent, of their home governments. Third, because the PMC is traded on the open market and is beholden to its shareholders, there exists greater potential for oversight, at least from a market standpoint. Fourth, while many PMCs retain the modern version of the Rolodex network, their structures are more permanent. Fifth, this permanence is reflected in the wider range of security related services they offer.

At this point, one may note that the modern private military company, while certainly a different entity from earlier mercenary organizations, does seem to closely resemble the mercantile companies of the seventeenth century. Nevertheless, as Carafano (2008) highlights, PMCs are different from their mercantile company ancestors in four areas:

1 The size and power of today's private sector is unprecedented.
2 Globally, in the post-Cold War era fewer business operations are owned and controlled by states.
3 Rules of global governance in the modern era define more effectively what type of behavior is permissible.
4 Today's global information environment (the press, the Internet, and so forth) allows for unprecedented transparency and activism in monitoring commercial activities.

(118)

Carafano, of course, believes that these four statements also apply to private military companies overall. Certainly, his observations reflect the changing nature of the international financial and political systems, but they may not prove the panacea for the large-scale embrace of privatized military. Private military companies may indeed be distinct from mercenary organizations, but this distinction alone in no way guarantees that these privatized military scions will be any less mercenary. The financial motivation still exists, albeit in a different manifestation, and because the financial motivation is now also business profit focused, some other problems may arise.

For now, these problems will have to wait. In the next section, I will turn to a brief discussion of the problem of categorizing the wide variety of security and military related services that private military companies fulfill. There are several different ways to categorize PMCs; for example, through their functions, by the services they provide, or through their proximity to the front lines. How one goes about categorizing private military companies can color her perspective when exploring ethical dimensions of using privatized force.

Private security companies and private military companies

In spite of acknowledging the distinction between private military companies and mercenaries, as well as the broad range of services these companies provide, some still lump these companies together. One author even refers to them as "Mercenary, Inc." (Silverstein and Burton-Rose 2000, xv). Many private military companies focus on offering military training and assistance to foreign allies; others even provide similar training to their state's own armed forces; some companies make it their business to collect and process intelligence; some provide the bulk of logistic support to military operations; while smaller companies are often hired for more sensitive, short-term contracts. In addition to the somewhat innocuous services, many of the private military companies offer armed security, and some of these have participated in combat in the past. In the 1990s, other PMCs, such as Executive Outcomes (EO) and Sandline, have been not only directly involved in combat but were also hired to do so.

The above list is merely a sample of the range of services provided by PMCs, but it serves to demonstrate the difficulty categorizing them, which leads to difficulty debating their permissibility, let alone a discussion of their accountability or regulation. This problem is compounded because many military companies perform more than one service simultaneously. While combat-focused PMCs, such as EO, seemingly disappeared by the end of the twentieth century, the 9/11 terrorist attacks on the United States and the ensuing war on terror offered wide-ranging, varied, and increasing opportunities for the private security industry.

As I introduced in Chapter 2, the use of private military companies has become pervasive in nearly every UN and NATO operation. For example, when I participated in the United States Army's Task Force XXI Army Warfighting Experiment (AWE) in March 1997, over 1,200 civilian contractors from forty-eight different vendors at the National Training Center in California, provided advice, maintenance, and technical support. Because technical support became such a vital component of military operations, some have suggested that the US military could not deploy without these private military companies. During the 2003 Iraq War, for example, PMCs provided support for the B-2 stealth bomber, the F-117 stealth fighter, Global Hawk UAV, U-2 reconnaissance aircraft, the M1 tank, the Apache helicopter, the Stryker, and many navy ships.

This widespread reliance on PMCs has not gone unnoticed in the international political community. The Montreux Document, released in 2008, was created to (1) acknowledge the widespread use of private military companies worldwide, and (2) to capture best practices and responsibilities suggested for private military companies themselves and their host and operating states. The Montreux Document introduction states:

> Private military and security companies (PMSCs) are nowadays often relied on in areas of armed conflict – by individuals, companies, and governments. They are contracted for a range of services, from the operation of weapon systems to the protection of diplomatic personnel. Recent years have seen an increase in the use of PMSCs, and with it the demand for a clarification of pertinent legal obligations under international humanitarian law and human rights law.
>
> (Maurer 2008)

In spite of the apparent progress of the Montreux Document to capture this privatized force phenomenon and attempt to regulate it, this document only provides recommendations and is not binding. Furthermore, it groups companies that provide these services into a broad category: "PMSCs." Thus, it is not as useful for our purposes because it does not have enough fine-grained resolution. One common sense way to categorize the PMCs, however, is by the type of support that they provide. We might bin PMCs into the following categories: operational support, military advice and training, and logistical support. But, we also need to account for other related services, including site security and intelligence. Unfortunately, as I will show in the following sections, categorizing by service type may not be ideal.

Differing ways to categorize private military companies

> These are firms that employ people who carry weapons to protect their clients and use them when necessary. Such firms are often labeled "private military contractors," although this more accurately refers to firms doing unarmed logistics work, such as KBR. PSCs are generally considered a subset of PMCs. Academics have spent years arguing over the appropriate terminology. I largely consider it an academic distinction that doesn't have much relevance to real-world discussion of the subject.
>
> (Isenberg 2009, ix)

As unambiguously pointed out by Isenberg, many discuss trying to categorize private military companies as a purely academic venture. Yet, I would argue that this "academic venture" is both relevant and essential for the debate over private force. What they fail to note is that, on the one hand, a clear distinction within PMCs would allow those companies that want to distinguish themselves from combat PMCs and mercenaries to do so based upon their categorization. For detractors of private military companies, alternatively, the categorization serves to distinguish different companies (and their services) to address the legal issues or whether a state should be privatizing specific activities. Consider the following. A logistics service company (e.g. Brown and Root Services, now KBR, Inc.) or a training services company, such as MPRI, may want to be distinguishable from companies such as Blackwater (Xe, now Academi), Aegis, or CACI. Blackwater and CACI offered different kinds of service, including escorting convoys, protecting civilian leadership, or even interrogating prisoners. These services are often difficult to distinguish from those that the military usually performs. For example, in its day, Blackwater was known for both protecting Ambassador Bremer in Iraq and guarding the streets of New Orleans after Hurricane Katrina, and in Iraq, Blackwater personnel were armed while KBR employees generally were not (per company policy).

Two camps: proximity to combat and service-based

There are basically two camps in the literature attempting to categorize private military companies; I will call these the proximity and the service-based methods, respectively. Although the two methods generally lead to the same results, both are attempts to clearly categorize what continues to be a complex and evolving industry, and both generally fall short.[8] It is worth briefly discussing each to point out their positive and negative attributes and how each method colors the larger debate. I will then introduce the method used for this project. Because we are focused on armed contractors, the problem of categorizing should be addressed first.

Proximity to combat categorization

Private military companies operate globally. While some of their activities seem mostly innocuous, such as supporting peacekeeping operations by providing convoy protection and guarding refugee camps, other aspects of their work are more controversial. As Silverstein and Burton-Rose (2000) reported:

> MPRI [trained] two Balkan armies and has won contracts in Africa and Eastern Europe. Vinnell trains the Saudi Arabian National Guard. Other private contractors support the Pentagon's own overseas military operations. Dyn-Corp, a billion-dollar company based in Virginia, has been especially active in supporting top-secret anti-drug actions in Latin America. A Florida firm called Betac [worked] closely with the Pentagon's Special Operations Command, which engages in covert activities in the Third World.
>
> (143)

Many of these companies operate overseas, and often, such as in the anti-drug operations, are operating directly in harm's way. One widely accepted method for categorizing PMCs, therefore, is proximity, like Singer's (2003) so-called "Tip of the Spear" typography. Simply put, PMCs can be categorized based upon how close they operate to the military's front lines—the tip of the spear. According to Singer, there are three types of PMCs in his typography: military provider firms, military consultant firms, and military support firms (100). Military provider firms are ones that participate in combat or directly support combat; military consultant forms provide training and advice to foreign military partners; and, military support forums provide logistics for ongoing operations and may be very specialized, such as mine removal operations. Singer further discusses the company he believes typifies each of these groupings: EO as a military provider firm, MPRI as a military consultant firm, and Brown and Root Services, his example of a military support firm, are discussed in greater detail.[9]

Others have adopted the tip of the spear typography and offered slight modifications. For example, Isenberg renames his versions of the three types of private military companies: military combatant companies, military consulting companies, and military support companies; although his company categories are essentially the same as Singer's, Isenberg (2009) uses "military combatant companies" to highlight those PMCs who have or are contracted for combat operations (25). Maintaining Isenberg's descriptions, one may note that many other PMCs do seem to fall neatly into the three categories; Sandline in addition to EO are examples of military combatant companies; Vinnell like MPRI is a consulting firm. DynCorp (in their support of LOGCAP), Titan, and Ronco are examples of military support companies. Yet, other PMCs are more difficult to clearly categorize. For example, where should CACI International fall? They are infamous for providing interrogators at Abu Ghraib, which seems to be close to the "tip of the spear." However, CACI was originally hired to update programming and databases for the Y2K transition. McIntyre and Weiss (2007) acknowledge this problem, they write,

> The companies engaged in a relatively benign facets of peacekeeping can also have activities that place them closer to the tip of the spear in other theatres (for example, a company could be engaged in supporting peacekeeping in Africa but undertaking different functions simultaneously in Iraq).
>
> (69)

For a company to be successful in this or any other industry, it must be dynamic and responsible to the changing needs of their employers or potential employers. A private military company that processes intelligence, for instance, may develop or acquire the ability to collect intelligence closer to "the front." It gains an ability to collect actionable intelligence, yet remains, for the most part, a military support firm. Similarly, for our purposes, we must ask where would one place a PMC like Aegis or Academi or ArmorGroup/G4S?[10]

Service-based categorization

Blackwater (now Academi) was a company that failed to match a tip of the spear typography. Attention has been focused on their armed security escort missions, and in 2014 the conviction of the employees involved in the 2007 Nasoor Square incident; however, Blackwater also maintained a robust military training service and multiple training facilities.[11] Perhaps then looking to the services these companies provide would be more fruitful. Categorizing PMCs by services provided or activities does not limit them to a static, frontline typography, which is proving irrelevant in a fluid, more global way of fighting wars, where there are no true frontlines. Today, PMCs undertake a range of services and activities that can be, for example, divided into Military Operational Support, Military Advice, Logistical Support, Security Services, and Crime Prevention services.[12]

While strikingly similar to a proximity typography, these categories focus on what the company provides for its principal. In the Military Operational Support category, therefore, PMCs provide support for, or participate in, military operations, but not necessarily in spatial proximity to the frontlines. Military Advice companies provide everything from weapons training for military forces, to tactics and force structure, to operational campaign planning. Logistical Support companies like KBR provide equipment, as well as establishing and running base camps downrange. Security and Crime Prevention services are typically manifested in the commercial market; they include guarding company assets and personnel. Examples of these types of services include multinational corporations turning to privatized protection, "especially in countries where governments exert little control over their territory. Here again, Africa is a proving ground, with private companies increasingly called upon to guard mining sites, oil fields, and other economic installations" (Silverstein and Burton-Rose 2000, 160). For our purposes, we might re-label these categories as:

Logistical Support,

Operational (or Technical) Support,

Military Advice and Training, and

Policing/Security.

We can now modify our definition of PMC to include the above service-based categories.

Private Military Company (PMC): C is a PMC if and only if C

1   a   is organized along corporate lines;
     b   is not integrated in a national force;
     c   operates from a business profit motivation;
     d   is a legal entity operating on the open market as its own entity or through a parent corporation;
     e   is organized into an established, permanent entity; and
     f   offers a range of military related services (e.g., training, advising, security);

2   operates in one or more of the following service-based categories:

     a   Logistical Support;
     b   Operational (or Technical) Support;
     c   Military Advice and Training; and
     d   Policing/Security.

This definition is more accurate, but it still does not capture a more basic and relevant distinction—whether the company (or its employees) uses force. Here one might also use the term "lethal," meaning the contracted right to use force (i.e. for self-defense or VIP protection) or, like EO, contracted for specific combat operations. While an armed-unarmed or lethal-nonlethal distinction might capture the concerns of employing armed contractors, it is only when combined with the service typography that one can capture the relevancy of being armed with the realization that (1) companies offer a variety of services, (2) individuals may work for different companies or their contracts may involve more than one service type, and (3) many but not all of these services require armed personnel or the use of force. While we are concerned primarily with the armed contractor phenomenon, we must acknowledge that the lines drawn between services are complicated and fluid.

PMCs in any of the four service-based categories might have armed employees, or a company like KBR, involved in running convoys of logistic supplies, may not have armed employees, but subcontract security from yet another PMC.

Nevertheless, the two categories that typically have an armed element are the Operational Support services and the Policing and Security services. Since companies that operate in the Operational Support category are employed for combat operations, much of the attention of the privatized military literature is focused on them. Operational Support PMCs are highly organized companies. They have a clear corporate structure to provide offensive combat operations, for a price. And, they can—like any PMC—be employed by states or other international actors. Kevin O'Brien (2007) elaborates:

> [These] companies are the ultimate evolution in capabilities and level-of-operation. They engage in military operations—across a spectrum where necessary, something most PSCs traditionally had not undertaken—and operate under contract in which their activities are designed to change the prevailing strategic environment in which they operate (such as defeating insurgency, ending a war, undertaking peacekeeping or peace-enforcement operations, rescuing a besieged government), especially in zones of conflict in the developing world.
>
> (38)[13]

It should be pointed out, however, that the notable examples of these companies— EO and Sandline—no longer formally exist. But, I mention them briefly below to demonstrate how these services were provided, to contrast them with the currently operational Policing and Security companies like Academi, Aegis, and Triple Canopy, and to show why lingering doubts exist toward armed contractors. Finally, while the Operational Support niche remains vacant, we may yet see another Executive Outcome in the future.

EO was a PMC based in South Africa that conducted operations in several African nations in the late-twentieth century. It was founded and manned mostly by former members of the South African Defense Force, including members of the infamous 32 Battalion, who found themselves highly trained and out of work as the units were disbanded at the end of the apartheid era. EO was involved in at least two major conflicts in Angola and Sierra Leon. EO is well known for contributing to the Angolan government's success in defeating UNITA and forcing them to a negotiated peace, and within a month of Sierra Leone's hiring of EO in 1995, Sierra Leone government forces had regained control of the diamond-rich Kono district. EO is controversial for both its use of hired former military and its supposed ties to Western diamond companies. It did not further their cause that in Angola, oil- and diamond-producing regions were the first areas secured by government forces trained by EO. Nevertheless, EO's actions in Angola and Sierra Leone were widely considered a success. The company's assistance brought the civil war to an end at a lower political and economic cost than could have been accomplished by these governments alone.

Sandline, alternatively, was an Operational Support military company established in London in 1990. Like EO, Sandline was operational in Africa, Sierra Leone and later in Liberia. The key executive in Sandline was Tim Spicer, formerly a British Army Lieutenant Colonel, who would later gain notoriety as the head of Aegis Defense Services, who had the $293 million contract corralling private security companies in Iraq, and in 2011 had a $400 million contract for assuming security operations at the US embassy in Kabul, Afghanistan. Sandline received the most notoriety for failed involvement in Papua New Guinea (PNG), where Spicer was even jailed for a time. In Sandline's PNG operation, the government was to grant Sandline mining concessions in its assistance in military operations to defeat rebel forces on Bougainville Island. Ironically, before the contract fell apart, Sandline subcontracted with EO for the operation. The PNG Army overthrew the PNG government over the "Sandline Affair," and when Spicer was released, he sued the PNG government for monies not paid. Other failures plagued Sandline. In Sierra Leone, Sandline was contracted to supply weapons and military services to the ousted government of President Kabbah. This government had been deposed by a military junta in alliance with the Revolutionary United Front (RUF), but Sandline's weapons shipments to the country were deemed in violation of the UN arms embargo.[14] Both EO and Sandline are no longer operational as corporate entities.

Opponents arguing against privatized force in general often draw parallels between Operational Support companies—or as others refer to them, "Combatant Companies" (Percy 2007a)—and companies providing Policing and Security. Both types of companies fulfill the definition of PMC; but Operational Support companies conduct offensive operations, and recalling the international anti-mercenary law, they seem to be contrary to the law. Supporters of Operational Support companies seem to lose much of the weight of their argument drawing distinctions between what they do as an Operational Support company and what a mercenary organization would do. It should be noted, however, that because EO and Sandline both "asserted that they would work only for sovereign states, and because of their incorporation into the structure of the armed forces of the state which hired them, they were not considered to be mercenaries" (Percy 2007b, 208) at that time.

Before moving on to discuss Policing and Security service companies, a few potential problem areas remain with Operational Support companies. First, despite the fact that there are currently no existing Operational Support companies, a precedence for their use has been established. Secondly, because of both their corporate nature and their expertise coincide within other types of PMCs, there is no guarantee against a future Operational Support company; the demand may materialize. Third, while there may be a generally accepted prohibition against the employ of an Operational Support company by Western states, this does not preclude other countries or multinational corporations from setting up or employing an Operational Support company. Fourth, it becomes increasingly difficult to categorize a corporate capacity for counterinsurgency, anti-terror, or other special operations, as many of these skills are also used in both Operational Support as well as Policing and Security companies.[15] Policing and Security PMCs (and even the other non-Operational Support PMCs) are similarly organized companies. They too offer military services, stopping short of combat. These services include protection and security services for a variety of employers, including states, NGOs, and other corporations. These companies often claim to use force only in self-defense; however, these services cannot be separated from the military context. Additionally, as I will later discuss in Chapter 5, there exists a fine line between self-defense and combat, as Blackwater and its Ass Monkey assets revealed in Najaf in 2004.

Private Security Companies (PSC)

Policing and Security service companies provided and continue to provide three main subcategories of services in Iraq and Afghanistan: personal security details for senior civilian officials, non-military and military site security (buildings and infrastructure), and non-military convoy security. While these operations undoubtedly include the carrying and periodic use of weapons, these services are distinct from those provided by the Operational Support companies. Nevertheless, as I alluded to earlier, these services and the companies that provide them are among the most controversial.

The Montreux Document tends to combine all private military companies under the heading of PMCS (or what we now refer to as PMC). It notes,

> "PMSCs" are private business entities that provide military and/or security services, irrespective of how they describe themselves. Military and security services include, in particular, armed guard in and protection of persons and objects, such as convoys, buildings and other places; maintenance and operation of weapon systems; prisoner detention; and advice to or training of local forces and security personnel. [And,] "Personnel of a PMSC" Are persons employed by, through direct hire under a contract with, a PMSC, including its employees and managers.
>
> (Maurer 2008, 6)

For this project, however, I will adopt the term "private security company (PSC)," as other authors have done, to focus on the companies and services that form the Policing and Security category. By their nature, PSCs are PMCs; they are PMCs of a certain sort, providing a more limited range of services, all involving the use or potential use of force. From the definition of PMC, we can extract a usable definition for "private *security* companies":

Private Security Company (PSC): C is a PSC if and only if C

1   a   is organized along corporate lines;
    b   is not integrated in a national force;
    c   operates from a business profit motivation;
    d   is a legal entity operating on the open market as its own entity or through a parent corporation;
    e   is organized into an established, permanent entity; and
    f   offers a range of military related, Policing/Security services.

Or, more succinctly, C is a PSC if and only if C is a PMC that provides Policing/Security services.[16]

Like their fellow PMCs, PSCs are organized on corporate lines (with boards of directors, shareholdings, and corporate structures), their work has clear contractual aims and obligations to their clients, and many of them now include capabilities in protection, intelligence, defensive, air and ground transportation skills. "While in an increasing number of cases—not just in Iraq but also in other parts of the world—PSC employees," Kevin O'Brien (2007) writes,

> are armed in defense of their asset (an installation, an individual, a piece of land, a population, or an NGO), the security and military skills are technical in nature and not [, unlike an Operational Support PMC's,] aimed at shifting the strategic landscape in which they operate beyond the immediate situations at hand.
>
> (38)

Furthermore, larger PMCs can and do have divisions of the company performing the policing and security services—PSC divisions.

Part of the controversy surrounding the hiring of PSCs concerns questions of whether the policing and security services that they provide are ones that the military ought to perform (i.e., the military is better able or the military has a duty to perform). These questions are not so easy to answer, so I will return to describe them in more detail later. Furthermore, looking at some ongoing operations, typically, PSCs are hired to provide (1) personal security detail (PSD), (2) site or installation security, (3) convoy security, and (4) policing. Including, of course, devising roles and training related to these services, but these also fall under a different category: Military Advisory and Training.[17]

In spite of a more accurate definition of PSCs, several difficulties remain. First, when armed and potentially using force, PSCs—while perhaps operating close to but not in actual combat—require technical skills that are identical to those used in combat. Often the effects of providing these services are certainly combat related, and the differences are often hard to tell. Often operations shift from self-defense and cross the line into offensive military operations, at least at the small unit level (a point I will discuss in Chapter 5 in detail). Certainly, to the employees under fire, the people they are hired to protect, their adversaries, and bystanders, there is no difference.

Second, often employees from one PSC may later work for another or even a wholly different PMC, blurring the line between mercenaries, PSCs, and PMCs. "Individuals who previously engaged in what could be defined as mercenary activity (that is, selling their skills on the open military market without corporate affiliation) might be employed by either a PMC or PSC" (O'Brien 2007, 39). Third, although PSCs do not meet the legal definition of mercenary, this does not entail that we should embrace their wholesale use either.

Finally, there remains the intuition that PSCs are wholly motivated by commercial success. This intuition remains grounded on the bias against a profit motivated force and is built upon the view, at least in modern, Western militaries, that not only is there a distinction between mercenary (as well as PMC and PSC) forces and so-called professional forces, but the later are preferable and superior. The next section seeks to elaborate on who these professional soldiers are.

The professional (or regular) soldier

The term "mercenary" has become so maligned that it is often used as a pejorative word to describe one's enemies, whether they are mercenary or not. Consider the Iraq war: Iraqi officials sometimes referred to American troops as mercenaries while Americans then countered, referring to Islamic volunteers fighting alongside Iraqis as mercenaries. As Percy (2007b) notes, mercenary can be used "to brand another group's soldiers and attempt to make them appear illegitimate" (50–1). How then might one show a distinction between mercenaries and professional soldiers, let alone PSCs?

Definition of a professional soldier

To answer this question, I must first define what I mean by a "professional soldier." One might answer, albeit naively, that professional soldiers are not mercenaries. In her defense, much historical literature relies on this simple distinction without further elaboration. The studies of ancient armies, as well as the writings of Machiavelli, Rousseau, and Kant, for example, extol the benefits of regular forces over the use of mercenaries. But then, one would counter that there are other types of soldiery: conscripts, militia, guerillas, and perhaps even child soldiers,[18] (although I think the later is a special case that I will let others explore), that are considered separate from mercenaries. The discussion then between mercenaries and professional soldiers and PMCs also involves some distinction between different kinds of soldiery. This is where this next discussion will begin. First, one characteristic that a professional soldier has concerns her technical warfighting skills.

Warfighting competence

In order to move toward the definition of a professional soldier, one must first differentiate between a regular soldier, a member of a standing army—which may include professional soldiers—and a member of the militia. Having the requisite military skills form part of the key argument for having a standing (or regular) military. The US Constitution itself captures the need for force that is versed in the art of war, and this need is so urgent for the state that the responsibility to "raise and maintain" is maintained at the national level. Students of military history know that debate raged over whether the United States should have a part time militia force or a standing army. Arguments for a militia force pointed to the legacy of the Minuteman and the fear over the use of a standing army by an oppressive regime. The proponents of a standing army, alternatively, saw the need for a well-trained and ready military to compete with other states.[19]

In order to achieve the requirements for a highly trained force, states could turn to mercenaries or develop a standing force of their own. Thus, one characteristic of a regular soldier is that she is trained in warfighting, and we should contrast this level of training with the training of militiamen. The militia fights as required; her main occupation lies elsewhere. Alternatively, the regular soldier's occupation is soldiery, period. Thus, we can say that a regular soldier:

> **Regular Soldier:** S is a regular soldier if and only if S's primary occupation is soldiery.

Nevertheless, while this distinction may eliminate militia from the discussion, both the mercenary's and the professional soldier's primary occupations are soldiery. This is of course correct; even in a conscripted army, the occupation of S is soldiery as long as she remains in uniform. It is also important to note here that both mercenaries and standing forces (including members of a professional military) fulfill the requirement for a full time competence in warfighting. Thus, a regular soldier could be defined as

> **Regular Soldier′:** S is a regular soldier if and only if S's primary occupation is soldiery, where S is either conscripted, impressed or volunteers into service.

Foreign status of mercenaries

From our earlier definition of mercenary, one characteristic is that a mercenary "is not a citizen or resident of the State in which she is fighting." Machiavelli (1995) drew from this characteristic to argue against mercenarism by "which a prince defends his state" (XII), and the foreign status of mercenaries shapes much of the historical debate over their employ, as discussed earlier. It was this same equating of the mercenaries with their foreign background that led Machiavelli to not only decry the use of mercenaries but to argue for a regular force comprised of the state's own citizens. One might be tempted to conclude, therefore, that regular soldiers would lack a foreign status. This, of course, is not always the case; regular armies may include soldiers from other countries, either impressed into service or as volunteers, and often do.

"Machiavelli's concerns about mercenary use revolve around a deep-seated feeling that native sons should fight for the republic," Percy (2007b) observes, "to ensure its health and success at war. Part of what made a republic strong was having its citizens fight for its preservation" (76).[20] The hiring of foreigners to fight negatively affects the strength of the state. Machiavelli maintained that the citizen had a duty to the state correlating to the flourishing conditions that a state provided its citizenry. He thought that "the republic is the common good; the citizen, directing all his actions toward that good, may be said to dedicate his life to the republic." To fulfill this duty in war, "the patriot warrior dedicates his death" (Pocock 1975, 201). The state, therefore, drew strength from its citizens' military service.

Machiavelli was not the first to see the hiring of mercenary force as a weakening of the citizen's civic duty and thus the weakening of the state. In ancient Greece, the tension between a citizen's duty to the state and a requirement for highly competent military force was present. Trundle (2004) writes that:

> The mercenary, however, challenged the community values of ancient Greek society because a mercenary was not a member of the community for which he fought and had no stake in that society, being neither citizen or landholder ... Mercenary service cut the links between citizens and community service, between a son and his household, between an independent farmer and his land, between the ideal amateur and professional specialist. Mercenaries cut the link between war and the political life of the community and thus the independence of the citizen who abrogated his responsibilities in needing a specialist to defend his home and his state.
>
> (2)

In spite of the Greek experience with mercenaries, as I discussed in Chapter 2, it was not until much later that widespread reliance on mercenaries diminished. For centuries, mercenaries were the traditional soldiers, and it has been relatively recently accepted that national ideals were tied to military service. The concept of a citizen's duty to the state, manifested through her military service, became a norm accepted by both citizens and the state governments.

Therefore, a second characteristic of the professional soldier emerges: the notion of the citizen soldier.

Citizen Soldier: S is a citizen soldier if and only if S is

1   is a citizen or resident of the state for which she is fighting, and
2   is integrated into a national force.

Of course, a citizen soldier could be a militia member or they could be a member of a standing force. For this discussion, we are concerned with the latter; thus, we can combine the definition of citizen soldier with the earlier definition of a regular soldier.

Regular-Citizen Soldier: S is a regular-citizen soldier if and only if

1   S's primary occupation is soldiery, and S
2   is a citizen or resident of the state for which she is fighting, and
3   is integrated into a national force.

We will turn to the question of financial motivation in the next section.

Motivation and the citizen soldier

Much has been made of the profit motivation of mercenaries, and it is the corporate manifestation of the profit motivation that causes some to question the hiring of

PMCs. Recall that both the definitions of mercenary and PSC (as well as PMCs in general) include this motivation characteristic. For the mercenary, she fights from a financial motivation; while the PMC entails three characteristics which all point to the corporate financial motivation: a PMC (a) is organized along corporate lines, and (c) operates from a business profit motivation, and (d) is a legal entity operating on the open market as its own entity or through a parent corporation.

Citizens, on the other hand, are supposedly motivated by their duty to the state. As some republican theorists like Machiavelli, Rousseau, and Voltaire point out, service to the state (in this case military service) is not only what is expected of each citizen, but also points to the motivation of a citizen soldier, namely the security of the state and themselves. Kant also draws distinction between mercenaries and citizen soldiers in his *Perpetual Peace*. For Kant, employing mercenary forces is simply "the practice of hiring men to kill or to be killed." He further argues that this practice of hiring men to kill or to be killed "seems to imply a use of them as mere machines and instruments in the hand of another (namely, the state) which cannot easily be reconciled with the right of humanity in our own person" (1983, 110).

Kant contrasts this conception with one of citizen soldiers. He writes that the "matter stands quite differently in the case of voluntary periodical military exercise on the part of citizens of the state, who thereby seek to secure themselves and their country against attack from without" (111). Kant views the motivation of mercenaries and citizen soldiers as distinct. Mercenaries are immorally motivated; they are instrumentally hired to kill by the state. Citizens, however, seeking to defend themselves and the state, are properly motivated, thus preferable as a force option. Yet, is Kant's distinction sufficient?

For a mercenary, having the primary financial motivation, or perhaps more importantly, lacking the civic duty motivation, has even further implications. First, mercenaries are free to choose to go to war. A regular-citizen soldier retains some obligation to fight when the state directs (at least in a just war). PSC contractors, like mercenaries, even when formed into units, can decide whether to join the unit. Their obligation, while contractual in the literal sense, is a business one. Mercenaries are not compelled to go to war; theirs is a war of choice.[21]

Nevertheless, there are of course historical examples of problems with citizen armies. First, as we have noted, if formed as militia, the citizen forces may lack the requisite training necessary to protect the state. Second, citizen armies have revolted, or even more relevantly, refused to fight, as the French did in the French Army Mutinies of 1917 (Keegan 1999). Third, one may question whether a civic duty is motivation enough to fight in an unpopular war, especially one that is protracted and where there are large numbers of casualties. Fourth, regular-citizen armies are paid some amount—perhaps not financially equivalent to mercenaries or PSC contractors, but they are compensated nonetheless.

We have already discussed and moved away from militia, and the potential that a regular-civilian army could revolt seems no different in terms of effects than if a mercenary army changed sides (perhaps following the money), as was previously discussed in Chapter 2. The third and fourth problems raise an interesting point: today's professional military are paid, and they do receive non-monetary compensation, such as leave, awards, and public recognition. However, it is the very notion of some type of compensation that also contributes to the arguments for volunteer forces over mandatory conscription.

Milton Friedman in his book, *Capitalism and* Freedom (1962), argued for volunteerism over conscription, and he made a compelling argument for a voluntary military to the *Gates Commission* that, in 1970, unanimously recommended an end to the military draft. A standing army had to be paid; a devotion to civic duty alone could not pay the bills or feed one's family. Friedman argued that volunteerism is superior to conscription based upon "economic efficiency: paying soldiers a competitive wage would result in being able to recruit and retain enough citizens to provide public security for their fellow citizens" (Carafano 2008, 32). Compensation is tied to military recruiting and retention and often may be strong motivation for an individual soldier. But, it is not the *raison d'être* of the soldier, nor what defines the regular-citizen soldier. I will return to this point in the next chapter.

For now, it is enough to include something about the motivation of regular-citizen soldiers in the definition.

Regular-Citizen Soldier: S is a regular-citizen soldier if and only if

1 S's primary occupation is soldiery,
2 and S

    a is a citizen or resident of the State in which she is fighting;
    b is integrated into a national force; and
    c is primarily motivated by her civic duty to the state (even if she also is financially compensated).

A defender of military privatization might be quick to point out that a PSC contractor, unlike a mercenary, could also be motivated by her civic duty to the state. She might further add that the civic duty motivation carries more weight than her financial motivation. For example, a recently retired soldier desiring to do something for the war effort finds that PSC employment provides the only alternative because she can no longer be a regular-citizen soldier. If you want to continue to fly and serve your country after you leave the military, then you could join a company like Blackwater's Ass Monkey. I know many other contractors who are very patriotic and see their employment as another way to serve their country. While I will discuss this objection in a later section, it is important to note that the employer of the PSC contractor is primarily business profit oriented, even if the PSC contractor is also motivated by a sense of duty to the state.

Oath of a professional

The discussion of the military as a profession has dominated recent military writings. Military theorists such as Huntington (1957), Millett (1977), and Janowitz (1960) have all attempted to capture the phenomenon of what it means to be a professional military soldier. That the US military acknowledges that it is a profession is widely accepted. For example, pre-commissioning education at the military academies and ROTC programs invest considerable time and energies on describing what it means for the military to be a profession and the duties and responsibilities that come with being a member of the military profession.

Although I will discuss the military as a profession in greater detail in Chapter 6, as well as the potential impacts of privatization on this profession, I need to briefly introduce the concept here to refine the definition of a "professional soldier." According to Millett (1977), all professions exhibit three characteristics:

1   specialized expertise attained by prolonged education and experience;
2   a responsibility to perform functions beneficial to society; and
3   a sense of professional corporateness.

Here, corporateness reflects "a collective self-consciousness that sets professionals apart from the rest of society." The specialized expertise in question here is of course war fighting, or as Millett puts it, "the management of violence." The professional soldier's responsibility is to provide national security. Millett writes that a "sense of corporateness flows from the educational process, the customs and traditions that develop within the profession, and the unique expertise and responsibility shared by group members" (Millett and Maslowski 1984, 133).[22] Part of what distinguishes the professional soldier from the regular-citizen soldier is the responsibility to provide for national security—a function beneficial to society—which is solely the responsibility of the professional soldier. In addition, the sense of professional corporateness reflects this responsibility. Here, professional corporateness is not merely a sense of belonging or membership. It is a sense of the responsibilities due to the society by the profession and responsibilities for ensuring the profession meets its obligations for the function society grants them; that is, the management of violence.

With this introduction to the military as a profession, I can now begin to construct what is meant by a "professional soldier":

Professional Soldier: S is a professional soldier if and only if

1   S's primary occupation is soldiery, a specialized expertise attained by prolonged education and experience,
2   and S

   a   is a citizen or resident of the State in which she is fighting;
   b   is integrated into a national force; and
   c   is primarily motivated by her civic duty to the state (even if she is also financially compensated); and
   d   is a member of a military profession, whereby the profession exhibits:

      i   a specialized expertise in warfighting;
      ii   a responsibility to provide national security; and
      iii a sense of professional corporateness.

There is another sense in which the professional soldier is different from the regular-citizen soldier, and even more so than a mercenary or a PSC contractor. The professional soldier swears an oath of allegiance when she joins the military. Some might argue, however, that the military oath is no different from many of the credos used by a variety of religious groups, social organizations, and more recently in the larger corporations. Even the International Stability Operations Association (ISOA), a volunteer association of PMCs with over 50 member companies that "provide a wide range of services in conflict and post-conflict environments," has a code of conduct (ISOA 2011).[23]

But, these codes are not only voluntary in terms of providing accountability, they do not reflect any obligations unless these obligations are clearly expressed in the contract. In the US military, the officer's commissioning oath tracks the duties and responsibilities from the president as commander in chief through the Department of Defense and the military chain of command. It is a government appointment. "A government appointment," Verkuil (2007) writes, "creates a public servant who, whether through the oath, the security clearance, the desire to achieve public goals, or the ... income of service, is different from those in the private sector. The office itself is honored" (1). Furthermore, when the president appoints military officers, these officers exercise command authority. In the US military, this delegation of the duties of an Officer of the United States occurs frequently; the term "officers" includes all officer ranks. He continues, "Ensigns or lieutenants are confirmed just like admirals and generals; they exercise significant authority when they carry out battlefield engagements" (103 and 108).

Note the US military officer's commissioning oath:

> I (insert name), having been appointed a (insert rank) in the US Army under the conditions indicated in this document, do accept such appointment and do solemnly swear (or affirm) that I will support and defend the Constitution of the United States against all enemies, foreign and domestic, that I will bear true faith and allegiance to the same; I take this obligation freely, without any mental reservation or purpose of evasion; and that I will well and faithfully discharge the duties of the office which I am about to enter, so help me God.

The professional soldier is of course bound by the law, but because she is a member of a profession—one whose corporateness include standards of moral and professional conduct—she operates in an environment where illegal behavior is unacceptable. "Duty, honor, country," the West Point motto, and other expressions of the military ethos" Carafano (2008) observes, "... demand honorable service for military members. These credos have real influence in ensuring that the vast preponderance of military personnel serve their nation well" (165).

We should now modify "professional soldier" to reflect the government appointment and oath taking characteristics to include the following: S

    e   swears an oath as a government appointee, embracing the duties and
        responsibilities as a member of the military profession.

Who is the professional soldier?

A professional-citizen army is closely linked to the current, normative concept of the state's identity. In the international community, having a strong military has and continues to be seen as vital for state survival. States, desiring to be taken seriously internationally, raised professional militaries made up of the state's own citizens. Recently, states who are considered morally good and successful are assumed to fight using their own citizens, and a professional military has become the norm for modern, Western militaries. Who is this professional soldier? The answer seems to be

Professional Soldier: S is a professional soldier if and only if

1   S's primary occupation is soldiery, a specialized expertise attained by prolonged education and experience;
2   and S

    a   is a citizen or resident of the State in which she is fighting;
    b   is integrated into a national force; and
    c   is primarily motivated by her civic duty to the state (even if she also is financially compensated);
    d   is a member of a military profession, whereby the profession exhibits:

        i   a specialized expertise in warfighting;
        ii   a responsibility to provide national security; and
        iii   a sense of professional corporateness; and

    e   swears an oath as a government appointee, embracing the duties and responsibilities as a member of the military profession.

The professional soldier and the PSC contractor

Although the use of professional militaries is a relatively recent evolution, the trend over the last century was the discarding of hired as well as conscripted forces and the adoption of military forces drawn from the citizenry of the state and formed as a military profession to serve the function of providing national security. Nevertheless, resurgence of recent privatized security activity represents a shift away from this post-Westphalian reliance upon citizen armies, where force remains in the public domain to provide security for the state and its people. Armed then with the definitions of a professional soldier as well as a PSC contractor, it is now possible to draw distinctions between the two types of military forces, and one can now compare whether one force or the other might be more suitable for serving a state's military needs.

On the one hand, critics of privatized force are quick to note that the use of PSC contractors, like their mercenary forbearers, appear problematic. PSC contractors make the decision to serve and potentially fight independently; it is a matter of choice. The PSC contractor cannot further claim a direct association with the cause for which she fights other than the contract to which she signs. Because a PSC might be (and empirically usually has been) external to a conflict, a PSC contractor may not possess the ideological instincts that motivates a volunteer or professional soldier. PSCs survive by being profitable.

Nevertheless, it is important to draw out two potential objections to the use of professional military over a privatized one. First, labeling of force as professional, even if the members of the profession meet the definition that we formed here, is no guarantee that the professional military will prosecute operations in an ethical manner; that is, in accordance with jus in bello. Second, professional soldiers are contractually obligated in a manner that may not be so different from the obligations of the PSC contractor. Certainly, by definition, the professional soldier is an expert in warfighting and a citizen of a state. Furthermore, as a member of the military profession, she serves to fulfill the function and the included responsibilities of providing national security, and she even swears an oath to the state as a reflection of her military obligations. Yet, in one telling historical example, members of a professional military, operating out of their perceived civic duty to their state, furthered the widespread destruction and loss of life enabling (if not directly participating in) the corrupt policy of the state government. The German military in World War II certainly perceived themselves as professional soldiers. They were citizens, integrated into the German military, and they were motivated by a sense of civic duty to the Nazi government. They were experts in warfighting; they drew upon a sense of professional corporateness; and, they even swore an oath as government appointees. Nevertheless, as history bore out, Hitler and other members of the Nazi regime were able to conduct expansionist wars and the extermination of millions.

Some may argue that had the German military truly been professional, the atrocities of the Second World War would not have occurred. But this argument seems wishful thinking. It is important to note that because the professional military serves a function of national security, a professional military could serve a corrupt state—professionally. Furthermore, war crimes may not be entirely linked to the type of forces a state uses. Isenberg (2009) recognizes this phenomenon. He writes:

> Mercenaries did not invent concentration camps, fire bomb cities from the air, use chemical or biological weapons, or use nuclear weapons on civilian cities. In fact, the bloodiest century in recorded human history was the twentieth, courtesy of regular military forces. Not even the most bloodthirsty mercenaries of centuries past could have imagined committing the kind of carnage that contemporary regular military forces routinely plan and train for.
>
> (6)

A professional military, however, when properly formed, provides a vehicle to conduct just war through the profession's standards of moral conduct. In other words, assuming the state desires just conduct in war, with a professional military, the state, as well as the profession itself, must link the civic duty due the state as well as the profession's sense of corporateness to include what is acceptable conduct in war.

The profession becomes the way for inculcating acceptable warfighting conduct. A privatized military would have to elucidate acceptable conduct in war and incorporate it as part of the contract, potentially overly complicating the actual contract.

Just as both professional soldiers and PSC contractors have a responsibility for just conduct of war, this responsibility is drawn in both cases from a form of contract: between a soldier in the state on the one hand and between the PSC contractor through the PSC to its employer—in this case the state—on the other. In the case of a professional soldier, the contract is manifested through the oath to the state; even more telling, each soldier literally signs a contract for her enlistment, and the oath in a way reinforces her commitment to the state. Additionally, like the soldier's contract, it is difficult to find any government activity that is not bound by a written contract. One might argue then that a professional soldier is also a contracted employee.

It seems clear, however, that these contracts are different, both in the way they are formed and in their scope. In the US military, for example, each soldier signs a contract with the state individually. In addition, the professional soldier falls under the Uniform Code of Military Justice (UCMJ), a legal code that regulates the conduct of soldiers. Currently, PSC contractors may or may not fall under portions of UCMJ depending on the contract and the employer; the PSC itself does not. Second, the enlistment contract that a soldier signs is between the US government and herself; there is no intermediary employer. The PSC contractor, alternatively, is employed by the PSC, which is in turn contracted with the government. Furthermore, corporate officers of the PSC are not government appointees like the officers of the military; their responsibilities are different. Therefore, while a contract plays a role in both the PSC's and the professional soldier's relationship to the state, each is different. I will return to the contract differences in the next chapter.

In this chapter, I have attempted to wade through the varied distinctions between the mercenary, the professional soldier, and the PSC contractor. The three groups are distinct. The Ass Monkey pilots were distinct from my soldiers; they belonged to a different organization; they were part of a PSC. But, they also had much in common. These pilots had been professional soldiers. They had specialized training and expertise; they remained citizens, and they had taken the same oath. On that day, it was clear that they were not primarily financially motivated. Furthermore, I argued that the modern day PSC is not the same as either the eighteenth century mercantile companies; nor, are its employees the same as mercenaries of the past in spite of their shared characteristics. The PSC contractor is distinct from the professional soldier. Yet, both can fulfill a state's requirement for forces. The questions become ones of cost: (1) Should force have a price? And (2), what of the other, non-monetary, costs that must be considered? I will turn to these questions next.

Notes

1 Blackwater pilots, Ass Monkey, also flew in support of the defense of the CPA compound in Najaf in 2004. See Prince (2014, 143–4).
2 There are of course some glaring issues with this simplistic formulation, but it gives us a starting point.

3 See also Doswald-Beck (2007, 120), Kinsey (2006, 19), and Trundle (2004, 22).
4 For greater detail on the problems of the Geneva Convention and mercenaries see Kinsey (2006, 19–21) and especially Percy (2007b, 169–79).
5 Tim Spicer (quoted in Kinsey 2006, 67). Spicer has a colorful and controversial history in the emerging private military scene. Often he himself is labeled as a mercenary. He began his military career in the British Army and later he was associated with EO. He eventually formed a group Sandline. As I will discuss later, Sandline is involved in Papua New Guinea and in Sierra Leone. In spite of being associated with these notorious private military adventures, he eventually formed the group Aegis Defense Services, which eventually won the contract in Iraq worth over $200 million to coordinate all private security company activity, and Aegis won the contract for embassy security in Kabul in 2011.
6 "Well for starters," Isenberg (2009) notes, "a majority of those working for private security contractors are Iraqi, and as such are nationals of a party to the conflict, so they don't qualify. Second, not all private security workers take direct part in hostilities" (7).
7 I would like to thank an anonymous commentator who recommended reinforcing the point that the semantic distinction, while important in some respects, was insufficient when comparing mercenaries and PMCs.
8 Another choice of PMC categorization is Percy's Spectrum of Private Violence (Percy 2007b, 59).
9 For background on the following PMCs, see MPRI (Avant 2005, 98–113, Singer 2003, 119–35; Silverstein and Burton-Rose, 169–75). MPRI is owned by L3, which has formed a new organization, Engility. "Engility was launched in 2012 as an independent company made up of leading businesses within L3's Government Services segments: including MPRI, Command & Control Systems and Software (C2S2), Global Security & Engineering Solutions (GS&ES), Linguist Operations & Technical Support (LOTS) and Engility Corporation and International Resources Group (IRG)" (http://www.engilitycorp.com/about/history/). Brown and Root/KBR (Singer 2003, 136–48). Vinnell Corporation, (Silverstein and Burton-Rose, 180–1). SAIC (Silverstein and Burton-Rose, 181). Booz Allen Hamilton (Silverstein and Burton-Rose, 181–2). DynCorp (Silverstein and Burton-Rose, 182–7). CACI International (Kinsey 2006, 101) Sandline (Avant 2005, 92–8; Kinsey 2006, 78–80). Executive Outcomes, (Avant 2005, 87–92; Kinsey 2006, 61–4; Silverstein and Burton-Rose, 164–6; Singer 2003, 101–19). Also see Madelaine Drohan's *Making a Killing* (2004) for in-depth discussion of both Sandline and EO as well as Alan Axelrod's *Mercenaries* (2014) for a history of private force and chapters devoted to these major PMC players.
10 G4S acquired ArmourGroup International in 2008 (http://www.g4s.com/en/).
11 See Academi's official website: www.academi.com. "ACADEMI is an elite security services provider. With expertise forged in the world's most challenging environments, our world-class network of security professionals design customized solutions for our clients to help them navigate the complex, sensitive environments in which they operate … Our cadre of experienced instructors create comprehensive training curriculums for both government and commercial clients that utilize our premiere 7,000 acre training facility located in Moyock, North Carolina." Also see Prince (2014).
12 Kinsey (2006) writes, "[Crime prevention] is not a military task but, because the PMCs and the private security companies (PSCs) are able to draw on an extensive network of retired military ensuing police officers, providing the service seems only sensible" (2).
13 O'Brien includes MPRI as a private military company for its support to the Croatian military in Operation Storm in 1995. (38)
14 See Avant (2005, 92–8) or Kinsey (2006, 78–80). Sandline's webpage shows the following message: "On 16 April 2004 Sandline International announced the closure of the company's operations. The general lack of governmental support for Private Military Companies willing to help end armed conflicts in places like Africa, in the absence of

effective international intervention, is the reason for this decision. Without such support, the ability of Sandline to make a positive difference in countries where there is widespread brutality and genocidal behaviour is materially diminished." (Sandline International)

15 For example, as Avant (2005) writes, "Beni Tal is an Israeli firm that advertises the capacity to carry out special operations and Control Risk Group, of the UK, specializes in crisis response and [had] a large presence working in post-war Iraq. US-based Blackwater also [fell into this nebulous] category. These companies promise to respond to crises offensively with armed personnel, but it is hard to know whether to call this a police/ SWAT-type action or a military special operations action. As their aim is combating not troops, but international criminal elements, they might be better characterized as internal security tasks[, those that belong to the Policing and Security services]" (21).

16 Other academics and authors use the term PSC, but I think it important here to define what the term entails to help eliminated some of the confusion over terminology. Of course, a PSC may also provide other services, but this again highlights the difficulty categorizing them or other PMCs.

17 Carmola (2010) says that PMSCs are protean in nature, and this is a good way of appreciating the complexity of understanding the private security field.

18 Cf., P. W. Singer's *Children at War* (2005a).

19 I am condensing this historical debate on militia versus regular forces greatly as the discussion focuses on not just regular forces but the professional soldier. For a more in-depth investigation into this debate see Millett and Maslowski's *For the Common Defense* (1984).

20 See also Millet (1977, 133 and 197).

21 But, the fact that a mercenary has only an individual responsibility for killing does not necessarily entail that a PSC employee is similarly restricted. On Percy's (2007b) view, even if the mercenary can't be sanctioned by the state, it might be the case that a company hired by the state to fulfill the security role be under the public sanction of killing on the state's behalf. So, as an employee of the PSC, she is not only individually responsible for her actions, she carries the state's responsibility to kill in its defense.

22 Cf. Snider, Nagel, and Pfaff (1999), Wakin (1979), and Spurlock (2001).

23 The ISOA (http://www.stability-operations.org) was formally the International Peace Operations Association.

4 Armed military privatization and the commodification of force[1]

> Weber's definition suggests that the state is constituted by its monopoly on the use of force. It is also, and perhaps more importantly, justified by its monopoly. This is what states are for; this is what they have to do before they do anything else—shut down the private wars, disarm the private armies, lock up the warlords. It is a very dangerous business to loosen the state's grip on the use of violence, to allow war to become anything other than a public responsibility.
>
> (Walzer 2008)

In his "Politik als Beruf," Max Weber (1926) famously theorized that the state "is a human community that (successfully) claims the monopoly of the legitimate use of physical force within a given territory."[2] While his proclamation, as I have discussed, is contentious (for example, the word "legitimate" is subject to controversy), his conception of a state's monopoly of force has had great influence in international politics. Of course, as many have pointed out, even if Weber is correct, a state can still delegate the use of force. The domestic analogies of police in towns or cities and sheriff departments' authority in unincorporated areas highlight forms of this delegation; perhaps even mall security guards reflect a form of this delegation. Furthermore, private security companies (PSCs) and their supporters seem to take this kind of delegation as support for the security missions that they provide. Simply put, if a state, S, can delegate the use of force, they might argue, then S can also delegate the use of force to a PSC. The contract formalizes this delegation, and the money exchanged is merely compensation for the service.

However, I think that this view is not only simplistic but it also both conflates the above domestic analogy and commodifies something that does not nor should not have a price—the use of force; at least this is what conventional arguments against commodification would argue. I contend, nevertheless, that the armed contractor phenomenon is a form of commodification, and it is this commodification of force that is distinct from and detrimental to a state's monopoly of force.[3] All things considered, states should not delegate force—force as a commodity—unless the consequences of doing so greatly outweigh the costs, monetarily and otherwise.

In this chapter, I will first define commodification and introduce Radin's (1996) conception of contested commodification. I will then discuss the commodification of force, and I will turn to two commonly held arguments against commodification in general. I show that neither will work to argue against the commodification of force (or any other commodification). Then, I will introduce a consequentialist argument against the commodification of force. Discussing the benefits and costs of employing PSCs will demonstrate that normally the hiring of PSCs is consequentially unsound. I will also address several objections to this view and show why they fall short. The argument against the commodification of force, manifested through the hiring of PSCs, negatively impacts Weber's monopoly of force. More precisely, the overall negative consequence of hiring PSCs should preclude their use unless in extreme emergencies.

What is commodification?

Commodification definition

The dictionary definition begins this brief inquiry about commodities. "Commodity" is defined as "an economic good," or "article of commerce," any "mass-produced unspecialized product," and, most inclusively as, "something useful or valued."[4] Karl Marx famously defined "commodity" as something that has both use value and exchange value. And, it is "an object outside us, a thing that by its properties satisfies human wants of some sort or another," ultimately tied to money and an impersonal market. F. Engels further added that "[t]o become a commodity a product must be transferred to another, it will serve as a use value, by means of an exchange" (quoted in Ertman and Williams 2005, 2–3).[5] For our purposes, then, commodities can be defined as goods or services that are (or could be) bought and sold. Conceptualizing these goods or services in this marketized way is known as commodification.

Therefore, we can define "commodification" as the process of something becoming understood as a commodity, as well as a "state of affairs once this has taken place."[6] What is most applicable here is how something, in this case the service, is then treated as a commodity—subject to the market, bought, sold, or traded for. A market exists whenever buyers and sellers exchange goods or services, where supply and demand determine price. This is the ideal market. Nevertheless, when "there are willing buyers and willing sellers, anything and everything will be in the market, regardless of legal rule forbidding the practice" (Ertman and Williams 2005, 2). Thus, I contend that (1) the market is not ideal, and (2) there are those who hold that some things should never be commodified.

Contested commodification

While commodification refers to how something comes to be conceived of as a commodity, as well as its resulting state of affairs, Radin's (1996) "contested commodification" more restrictively (and relevantly to this chapter) refers to "instances in which we experience personal and social conflict about the process and the result" (xi). Contested commodification, for example, applies when one intuits that prostitution is a wrongful buying and selling of one's sexuality. Sex as a commodity places it within the effects of market influence. Radin further suggests a continuum "stretching from universal noncommodification (nothing in markets) to universal commodification (everything in markets)" (xiii). (I think that the actual market lies somewhere in between.) For this project, I will be focusing on the conception of force as a contested commodity. However, for simplification, I will use the word "commodification" in this following way. By commodification I mean:

Commodification: Assigning price to something (a kind of service in this case) that should not have a price.

Much of the recent literature on commodification centers on prostitution or medical and legal issues surrounding the use and ownership of body tissue (e.g., ovum and sperm) and organs.[7] While I will suspend my judgment on either debate, I find the debates useful to highlight the intuitions that those arguing against prostitution or commodification of organs seem to hold:

1   some things ought not be for sale, and
2   these things ought not to be treated as if for sale.

Thus, I take commodification to have normative weight, not merely descriptive, and I contend that the use of force may be one of these things that ought not be for sale.

Commodity of force

Consider a contemporary example for illustration. While much of the attention remained focused on US foreign policy surrounding the wars in Iraq or Afghanistan, an obscure conflict, one that has existed for over ten years, has attracted some international attention. Although piracy on the Somalia coast has been ongoing, the recent seizures by pirates of a vessel containing tanks and other armaments and another with an estimated $100 million in oil have drawn the ships of several navies to converge on that area. This has resulted in a version of a Mexican standoff in the Somali waters.

Not historically being a company to pass on opportunities in the emerging security arena, Blackwater (now Academi) offered in 2008 to provide security for merchant ships—security for a price—and their proposal has been informally endorsed by some US Navy personnel (Houreld 2008). According to Blackwater's (Academi's) statements,

Blackwater Worldwide today announced that its 183-foot ship, the McArthur, stands ready to assist the shipping industry as it struggles with the increasing problem of piracy in [Somalia's] Gulf of Aden . . . . As a company founded and run by former Navy SEALs, with a 50,000-person database of former military and law enforcement professionals, Blackwater is uniquely positioned to assist the shipping industry.

(Blackwater Worldwide 2008)

Generalizing and putting their proposal in argument form might look like the following:

Argument for Commodification of Force—PSCs:

P1. Force is a Commodity.
P2. States have a monopoly on use of force. (ala Max Weber)
P3. States' requirements for use of force can be satisfied by either

    a   regular/standing military;
    b   conscripted military;
    c   non-conventional forces (e.g., impressed sailors, child soldiers, etc.);
    d   hired/contracted forces; or
    e   a combination of the above.

P4. (P3d) can be satisfied by

    a   mercenary individuals or units; or
    b   private security companies.

P5. States' requirements for use of force is often perceived to be under or unfulfilled. (This can be due to new threats, multiple resource taxing operations, changing missions, etc.)
P6. Thus, states can either

    a   increase the regular military;
    b   institute or expand conscription;
    c   resort to use of non-conventional forces; or
    d   contract force (through PSCs).

C7. P6 a, b, and c are politically untenable, and employing mercenaries and mercenary activities themselves are illegal (as was discussed in Chapter 3). Thus, states will contract force.

Does this sound farfetched? Erik Prince's recent call to use contracted forces as "boots on the ground" versus the Islamic State in Iraq and Syria (ISIS) illustrates this line of thinking (Lamothe 2014).[8]

An objector to the use of PSCs might look at the argument and say, "So what?" The argument, if successful, merely shows that states may contract force. It does not argue for its moral permissibility. And I agree, but what the argument shows is that there are both political and pragmatic reasons that a state might have for hiring armed contractors, and often arguments of this type are construed as having moral weight. Nevertheless, there is deeper concern here. While one might grant the PSC proponent premises 2 through 7, some would argue that force is not a commodity; P1 is false. Alternatively, as I will show, force may in fact be a commodity, but it should not be in normal circumstances. In other words, because a state can draw upon hired PSC forces to satisfy its requirements for use of force, does not mean that it should.

Force commodification and the Weberian state

As discussed in Chapter 3, force has been a commodity in the past, and as Avant (2005) notes, the demand for PSCs has bloomed again. She writes that

> Concomitant with the increase in supply was an increase in the demand for military skills on the private market—and western states that had downsized their militaries, from countries seeking to upgrade and westernize their militaries as a way of demonstrating credentials for entering into western institutions, from rulers of weak or failed states no longer propped up by superpower patrons, and from non-state actors such as private firms, INGOs, and groups of citizens in the territories of weak or failed states.
>
> (31)[9]

The numbers of security contracts seems to show that (1) companies do bid on the contracts, (2) the value of these contracts are impacted by market forces, and (3) these security contracts function like other commodified products and services. Force, therefore, seems to be treated as a commodity, but the question at hand is should it be.

As discussed in the last chapter, the commodification of force represents an extensive shift in Western state policy. Since Would War II, for example, debate seemed to focus on whether forces should be conscripted or made up of volunteers. Although the "production of the goods needed to wage war long ago became the domain of the market ... [,]" Singer (2003) writes, "the service side of war was understood to be the sole domain of [states]" (7). Providing for national defense remains one of the essential tasks of states, and according to Weber, it defined what a state was supposed to be. Moreover, the international system of states seems to be built upon this seemingly fundamental concept, dating back before Kant. Enlightenment thinkers like Kant thought war to be something bad, causing more harm than good. Kant (Kant and Humphrey 1983) writes in *Perpetual Peace* that: "[war] makes more evil people than it takes away." Kant, therefore, "lays out a set of conditions that must be met for peace to be possible, one being that states have well-formed constitutions and another being that states participate in a federation of states ..." (Challans 2007, 150).

Recall that in the international system, states alone may organize legitimate violence; "other groups (such as international organizations, private citizens, and non-state actors) may use force only with their permission. States, in other words," as political theorist Cockayne (2007) affirms, "are the mutually recognizing, oligopolistic principals, each with a monopoly on legitimate violence within their own territory" (199).[10] The increasing reliance on PSCs calls these norms into question and perhaps leads to a kind of contradiction. Can the use of PSCs coincide with the existing international system and the state?

Force is commodified other than monetarily

There is another way in which force may be commodified that need not be financial. In other words, one might be compensated for using force, or force might be incentivised. Militaries often use other reward systems for bravery and good conduct, for example, medals, badges, rank, or leave. Napoleon famously said, "A soldier will fight long and hard for a bit of colored ribbon." This quote and the other non-financial rewards or gains highlight the psychological observation that soldiers fight for other than financial reasons; in Napoleon's case, he is alluding to fighting for personal glory rewarded by military decoration. The reasons why humans fight are varied, and the study of this phenomenon is complex.[11] The problem with this observation is that it conflates individual motivation for completing a task and the psychological reasons for taking up arms with assigning a non-monetary value to force. I think that this line of argument does serve to provide useful insight into the questions of why people become mercenaries (perhaps even why these PSCs are mercenary); nevertheless, it merely suggests that there are other conditions of value which may or may not entail commodification in general, and in this case the commodification of force.

In addition, for states and even multinational companies, the hiring of PSCs might not be an intentional commodification of force. Rather PSCs could merely provide an alternative to the need for security. A need to protect investments abroad increases demand for security.[12]

Thus, while not intentionally turning force into a commodity, they are making it one nonetheless. One way to argue against the hiring of PSCs is to argue that force should not become commodified. In the next section, I will look at two arguments against commodification to see whether either works as an argument against commodification of force.

Two arguments against commodification and why they fail

As Carol M. Rose (2005) points out in "Whether Commodification?," "[o]ne of the major contributions of Radin's classic analysis of commodification was her refinement of the concept of inalienability." According to Radin, Rose writes, "there is an ideal form of inalienability in which the goods or services at issue are never allowed to be transferred" (408).[13] One might capture Radin's idea with the following: a good may have a property, inalienability *(i)*, such that anything with '*i*' cannot be exchanged.

Some may then add that the use of force is one of these inalienable services. Considered another way, the "use of force" has some property, *p*. The state has the monopoly on *p*. And, in accordance with Radin's view, *p* has some special standing. While one may intuit that *p* should never be transferred at all, where is the argument? Those who argue against commodification usually go one of two routes. In order to form these arguments, it is worthwhile to briefly look at two analogies—topics that are often at center stage in commodification discourse—prostitution and organ harvesting.

The prostitution analogy

Sandel (1998) succinctly captures two common objections to prostitution. Sandel asks the reader to first consider that one might "object to prostitution on the grounds that it is rarely, if ever, truly voluntary. According to this argument, those who sell their bodies for sex are typically coerced, whether by poverty, drug addiction, or other unfortunate life circumstances."

In this first objection, prostitution is considered wrong because sex is marketized involuntarily. Sandel continues by describing a different objection to prostitution. In this second objection, prostitution is considered wrong because "prostitution is intrinsically degrading, a corruption of the moral worth of human sexuality." Prostitution is morally wrong on this view "even in a society without poverty and despair, even in cases of wealthy prostitutes who like to work and freely choose it." I will call the two objections the consent objection and the degradation objection, respectively.

The consent objection does not really apply for our purposes. True, one might be coerced by poverty or other unfortunate circumstances to join a PSC, but this puts one's employment into question, not the commodity of force. An alternative view might hold that a state has no choice but to hire a PSC. While this may seem rather fanciful in the case of large states, it could be the case for smaller or failing states. However, I think this concern can be answered by my argument, which I will soon present. Since the consent objection would not apply, I will turn to the degradation objection.

The degradation objection, however, may be applicable and is certainly used in the commodification literature.[14] As Ann Lucas (2005) observes, "most objections to prostitution are commodification based. Some argue, for example, that exchanging sex for money degrades a human attribute, sexuality, the debasement of which impedes human flourishing" (248). As Sandel (1998) puts it, prostitution is "a corruption of the moral worth of human sexuality."

The bioethics analogy

Bioethicists are often faced with a similar argument against the harvest of human organs. Exchanging organs for money degrades human dignity because the organ was initially part of the body. Because it was part of the body, marketizing the organ debases the dignity of the body and degrades the person's dignity. While perhaps *prima facie* convincing, the argument is of course circular, at least as I have described it. To work, this kind of argumentation would need to be far more complex. Further, it relies on notions of bodily integrity and human dignity, which seem imprecise concepts at best, ones that need further argumentation. (I will return to this problem momentarily.) From these two analogies, however, we can form and discuss the first argument versus commodification of force, that marketizing force corrupts an ideal, and the second, that marketizing force undermines the state.

Argument 1: marketizing force corrupts an ideal

What I have in mind for the first argument is something along the lines of Sandel's military service commodification objection. Sandel (1998) argues that, "[a]n excessive role for market corrupts an ideal the practices properly expressed in advance—namely, the ideal of citizenship as the republican tradition conceives it." As discussed in Chapter 3, citizens are supposedly motivated by their duty to the state. As some republican theorists like Machiavelli and Rousseau point out, service to the state (in this case military service) is not only what is expected of each citizen but also points to the motivation of a citizen soldier, namely the security of the state. This is the ideal of citizenship to which Sandel refers.[15]

The argument against the corruption of this ideal, in this case military service, includes that there are certain practices that in some way reinforce the ideal. For example, the concept of a militiaman dropping his work to bear arms for defense of the state would seem, at least according to Sandel, to manifest the ideal of citizenship. Conscription, on the other hand, (or even paying a professional soldier on Sandel's account) is a practice that takes something away from the ideal of citizenship; it corrupts it. As Posner (2005) points out, "Conscription is a form of slavery, and slavery is the ultimate commodification. Conscription treats the persons conscripted as if the state *did* own them" (130, with original emphasis). Therefore, the argument would conclude, conscription is wrong because the practice corrupts an ideal that the practices properly expressed in advance.

Applied as an argument against the commodification of force, the hiring of PSCs (or mercenaries) is a practice that also corrupts the republican ideal of citizenship. While seemingly forceful, this argument against commodification of force suffers several problems. First, a defender of this view would need to argue for the assumption of an ideal of citizenship. While I also believe that there exists an ideal of citizenship, which I will be discussing in greater detail in Chapter 6, I do not think that the corruption of the ideal of citizenship alone provides sufficient grounds for argument against the commodification of force. Second, this argument must demonstrate how commodified force performs the corrupting function. In order to make this move, the defender must bootstrap the conception that any time force is treated as a commodity, the ideal of citizenship somehow breaks down.

Third, Sandel (1998) uses the argument against commodification of military service to imply that even a professional force, raised from a state's own citizens would be a kind of commodification. He argues that all citizens have an obligation to serve and this "is not the sort of thing that people should be free to buy or sell. To turn such service into a commodity—a job for pay—is to corrupt or degrade the sense of civic virtue that properly attends it." I disagree of course; as I discussed in the last chapter, standing armies have to be paid; a devotion to civic duty alone does not pay one's bills. Monetary compensation does not need to equate to commodification.

Finally, it seems to me that in extreme circumstances, perhaps for example when standing forces are insufficient for impending attack against the home state, that hiring other forces, perhaps PSCs, would better enable the state and her citizens to defend themselves. In other words, it is not the commodification of force that corrupts the ideal of citizenship; as I show in Chapter 6, it is the state's reliance on other forces such as PSCs that undermines the relationship between the military profession and the citizens of the state, even impacting the jus ad bellum decision to go to war, that corrupts the ideal. Therefore, the argument that marketizing force corrupts an ideal does not work.

Argument 2: marketizing force undermines the state

Radin (1996) argues that commodifying of things intimately associated with personhood, such as "work, sexuality, wisdom, character, and bodily integrity," inhibits human flourishing. Commenting on Radin, Lucas (2005) observes that

> Putting such personal attributes and attachments on the market converts them into possessions, "fungible objects" considered separable from individual, when in fact to detach them degrades or harms the person. Thus, prostitution—commodified sex—is detrimental to humanity because sexuality is integral to person and should be neither traded on the market nor analyzed in market rhetoric.
>
> (249)

Likewise, an argument that marketizing force undermines the state purports that the authority over the use of force is considered integral to states and also "should be neither traded on the market nor analyzed in market rhetoric."

If we hold that Weber is correct—that the state has a monopoly on the use of force—then commodifying the use of force makes it in some way separable from the state. Adopting this line of thought, then, one might argue that the use of force should not be commodified. But, before one rushes to adopt this argument, let us look at the consequences of doing so. First, one might imagine, if the use of force is indeed integral to the state and was "sold off," the state would not seem to be the same. Put another way, the state, S, as Weber knew it, would no longer exist. Another, perhaps S', would take its place. Setting aside the possibility that S' is preferable to S, an argument along these lines must account for this change.

Second, the argument that the commodification of force somehow undermines the state also suffers similar difficulties that plague those who argue against the commodification of human organs because they are integral to a person, and marketizing anything integral to a person degrades personhood and human flourishing. I do agree that the state is the primary principal in the international system and that the state should have the monopoly of force. It is not clear, however, that the use of force includes the concept of inalienability that the arguments against organ transplant above include.

Nor, is it clear that the hiring of forces to serve the function of the state's use of force degrades or eliminates the state's monopoly. As was discussed at the beginning of this chapter, states often delegate the use of force. However, a person using this line of argumentation, it seems, would also have to accept that even these other seemingly innocuous forms of delegation of force would also undermine the state. On the one hand, delegating force seems to introduce concerns with control over those forces (and there certainly have been issues of control over the conduct of PSCs), but to me it seems that the monopoly of the use of force in these cases does not disappear. Rather the monopoly is manifested in a different way.

One might counter that the hiring of other forces to fill the function of the state's monopoly on the use of force merely restricts the ways in which force can be commodified. For example, one might argue that force can never be a commodity on the open market. I would agree to the basis of this qualification. However, accepting this restricted view also means accepting that force is no longer inalienable. I would agree that there are circumstances (extreme ones) where a state might hire PSCs, and I concur that hiring PSCs could undermine the state. What is not clear is whether the commodification of force is always wrong and should never be accepted based upon its inalienable inseparability from the state.

Why force should not be commodified

Thus far, I have discussed two arguments against commodifying force and why they do not work. In this section, I will explore a different—consequentialist—argument against the commodification of force through the hiring of PSCs. I do this for two reasons. First, a consequentialist approach avoids the problems associated with the aforementioned, seemingly absolutist arguments. Second, defenders of the use of PSCs readily point to their financial benefits and expected increase in efficiency as the starting point for their argumentation. On my account, if the benefits really do outweigh the foreseeable and expected costs, then hiring PSCs may, in that case, be a morally viable option.

However, I contend and will show that, unless we institute broad contractual control and oversight reform, unless we truly understand the costs and benefits, we should have a standing, *prima facie* prohibition against employing PSCs. What is required then is a close look at the very argument PSC supporters use—that privatization of certain functions, like other forms of governmental outsourcing, result in both cost savings and increased efficiency.

The benefits of hiring PSCs

Privatization in its many forms has been historically "associated with comparative advantage and competition," Avant (2005) writes, "leading to efficient and effective market responses and contrasted with staid, expensive, and backward-looking bureaucratic response" (35). Thus, the two readily apparent areas that highlight the benefits of hiring PSCs, and outsourcing in general, are in cost savings and in increased efficiency. Both these kinds of benefits can be viewed in the short term and in the longer term. In addition, hiring of PSCs leads to other benefits, many of which fall outside of the financial and efficiency categories.[16]

The apparent sudden hiring of PMCs including PSCs did not happen overnight. As discussed in Chapter 3, the increased use of PMCs began in earnest with the corresponding privatization programs occurring in the late-1980s. Market-based approaches began to emerge in various government agencies, such as with the Internal Revenue Service (IRS), and also began to become popular in business, utilities, and other industries. With other government departments outsourcing differing requirements, it also made sense that the military was also considered. Indeed, the military seemed ideally suited to some for privatization for at least some functions because of the broad range of varied missions and the limited times the military was called upon. This led to the so-called "Third Wave" privatization initiative within the Pentagon.[17] Isenberg (2009) observes that many thought it "more efficient for the military to call on a group of temporary, highly trained experts in times of war—even if that meant paying them a premium—rather than to rely on permanent standing army that drained resources (with pension plans, health insurance and so forth) in times of peace" (15).

Hiring contractors seems to provide significant operational benefits to the Defense Department. First, using PMCs to perform non-combat activities frees up soldiers to perform combat missions. Second, it takes a relatively long time for the military to develop a new capacity (e.g., information warfare), and PMCs can both be hired faster, thereby enabling the military to rapidly adopt a new capability, and can be deployed quickly to provide these critical support capabilities. Third, PMCs are also able to provide expertise in specialized fields that the military may lack or not possess in sufficient numbers. A prime example of this is linguist capability (Schwartz 2009, 2).[18]

One of the most concrete benefits includes the saving of government spending. Schwartz writes: "[PMCs] can be hired when a particular need arises and be let go when their services are no longer needed. Hiring contractors only as needed can be cheaper in the long run than maintaining a permanent in-house capability" (2). In fact, a report from the 1995 Defense Science Board "suggested that the Pentagon could save up to $12 billion annually by 2002 if it contracted out all support functions except actual war fighting" (Isenberg 2009, 2). A clear example of how the US Army sought to capitalize on privatization is in logistics. The Logistics Civil Augmentation Program (LOGCAP) is an Army program established in 1985 to use civilian contractors in wartime and other contingencies, such as in Bosnia and Kosovo.[19]

As a fourth benefit, an international hiring of multinational PMCs has, in some respects, aided in the US government's foreign security assistance programs. One might also add that in places such as Iraq and Afghanistan, the US Defense Department has relied on PMCs recruited internationally, employing over 242,657 contractors in 2009 in the CENTCOM region, providing a much larger labor pool than could be provided by the military alone.[20] When a mission requires a long term commitment, such as with training other militaries, PMCs can provide greater stability in such programs, where the same personnel stay beyond a short deployment.

PSCs in particular are widely used in developing countries, where non-state actors such as multinational companies are increasingly reliant upon such PSCs to protect their employees and assets. In addition, charities and other NGOs also use PSCs to train their own employees in survival skills and to directly protect them on the job. International groups, aid organizations, other states, regional coalitions, and even the UN are turning to PSCs to provide security and PMCs for other support during humanitarian missions. These organizations seem to acknowledge that resolving humanitarian crises may be beyond the capability of any single state. Even the association of PMCs who make up the ISOA advertise their focus on supporting humanitarian operations.[21] PMCs initially seem ideal to support these humanitarian missions. The flexible nature of PMCs makes them seemingly well suited. With the increased hiring of PMCs by countries such as the US and Britain, as well as the seemingly unending need for support of peace and security missions, the PMC industry is seen by many as a growth sector. In September 2005, the Stockholm International Peace Research Institute predicted that the industry was likely to double in size in five years, with industry revenues of $200 billion in 2010 (Perlo–Freeman and Sköns 2008).[22]

Hiring PMCs, although widely accepted and utilized, is controversial, but employing PSCs seems to draw the most attention. While much of the literature on the use of PSCs seems negative, supporters of the use of PSCs, such as Carafano (2008), argue "There is no reason that the government should not exploit the advantages of free trade and outsourcing ... if these practices deliver the best value for service" (126). Nevertheless, in his enthusiastic endorsement of PSCs, Carafano also reveals some of the potential pitfalls. First, do these benefits apply in every circumstance of PSC employment? Second, while in "the best circumstances these contracts can enhance military effectiveness only with minor functional losses (such as increased costs)" (Avant 2005, 63), war is one of the most chaotic environments. Can reality match expectations?

The US military could not conduct the varied operations in Iraq and Afghanistan without the hiring of PMCs and PSCs.[23] Nevertheless, (1) the need for these other companies was not foreseen, and (2) the reality on the ground, combined with perhaps some miscalculation in terms of both the scope of the task of rebuilding Iraq and the appearance of the insurgency, suggest that perhaps the hiring of PSCs was facilitated by the conflict itself. "Since US forces were not available to protect those doing reconstruction work, such firms had no choice but to turn to [PSCs] to protect employees," Isenberg (2009) writes. "Put another way, while [PSCs] provide valuable services in Iraq, monumentally poor planning created the need for them" (157). In the economics literature, questions of outsourcing certain functions revolve around questions of the "efficiency of contracting." While PSCs can provide many benefits in the near term, with the changing nature of warfare—especially in counterinsurgencies and today's complex operational environments—marginal costs often begin to increase.

The costs of hiring PSCs

> [Outsourcing to PMCs provides] a tidy picture: the Army becomes a lean, mean killing machine, while civilians peel the potatoes and clean the latrines. But there's a reason that companies like General Motors existed in the first place. Effective as outsourcing can be, doing things in-house is often easier and quicker. You avoid the expense and hassle of haggling, and retain operational reliability and control, which is especially important to the military.
>
> (Surowiecki 2004)

Much of the focus on the hiring of PMCs falls into either the financial savings or the increased efficiency that outsourcing would generate. Recall that these two reasons helped propel the large-scale privatization initiatives in government. In the last section, I focused on the many benefits that hiring PSCs could generate. In this section, I will turn to the costs associated with the hiring of PSCs. There are certainly negative consequences of hiring private security in terms of financial costs and operational efficiency; but, there are additional consequences that are associated with this commodification of force. Moreover, discussions of these other "costs" are often overshadowed by debates of cost savings versus efficiency, as well as legality and control.

Hiring PSCs incurs secondary costs, including personnel and contract replacement, workers compensation, increased insurance premiums, evacuation and rescue costs, and increased reconstruction costs.[24] Especially when considering the total costs of contract fulfillment, one must question whether outsourcing results in actual, promised cost savings. A Deloitte & Touche survey of 1,500 chief executives suggests that either the savings are overestimated or not transparent. The survey notes, "only 31 percent believed that outsourcing had generated significant savings, and 69 percent were disappointed in the overall outsourcing results" (Singer 2003, 157). While these results may not be statistically alarming, they do suggest that even when financial costs can be compared, the savings may not exist.

In addition, financial costs alone may not be a sufficient measurement of successful outsourcing. Recall that in the business world one tries to achieve the three successful characteristics of cheap, fast, and good; yet, it seems that most can only achieve two of the three. Likewise, increased efficiency usually has a price. For example, "[logistics] officers often talk about value in terms of cost/speed of delivery/quality of service. If you need it tomorrow in the war zone, you can't expect Federal Express to get it there for you" (Isenberg 2009, 22).

These business dilemmas, particularly when one is discussing efficiency, are magnified in both the spheres of domestic politics and in international relations. As Verkuil (2007) writes, "democracy is not defined by efficiency alone. Accountability is a countervailing principle of democracy. It may but need not be efficient. Indeed, as the Supreme Court has recognized, the Constitution is sometimes intentionally 'inefficient'" (4). Whenever efficiency outweighs accountability, Verkuil contends, the possibility exists of efficiency undermining democratic values.

These costs are harder to quantify but must be considered. In the debate over the hiring of PSCs, if one merely looks at a make-or-buy choice, then the argument might be only a function of whether a PSC could perform the requirement better and at lower cost—a transactions cost approach (66). However, as will become apparent, hiring of PSCs is not a mere transaction cost evaluation.

In order to better discuss the negative consequences of hiring PSCs, I will categorize them generally into four areas. The first category is what I call contracting costs or *the cost of doing business*. This category includes issues arising from contracting uncertainty, issues of traction and repeated contracts, the lack of an ideal marketplace, defining contractual success, dependency on private security, walk-outs and strikes, overcharging, and weak oversight leading to wasted money. The second category, *control and security contracting*, includes concerns of subcontracting, multiple and transnational stakeholders, reliance on third country nationals, "independent contractors," shifting companies, and the question of guarding the guards. The third category introduces the problems associated with PSC agency, the evolving world order, and when the primacy of security conflicts with a competitive market. This category is called *private security and the changing international landscape*. The last category discusses when hiring private security involves a conflict between *the public good and the private good*. In this category, I will look at the effects of hiring PSCs on alliances and waging counterinsurgency, as well as the international reputation of the hiring government. I will first turn to contracting costs.

Contracting costs: the costs of doing (government) business

Contracting security requires some financial costs, often comparable to government expenditures on security. Nevertheless, security contracting incurs unique contracting costs. Furthermore, contracting security, particularly in a conflict environment, creates other costs not typically found in other outsourcing.

Contractual uncertainty

Government contracts come in a variety of flavors—from the most common, fixed-price contracts[25] to the more notorious ones, such as cost-plus contracts and sole-source contracts. In terms of transaction cost comparison, a fixed-price contract is preferable, as the agent (contracting company) receives set fees for the services the contractor provides. Cost efficiency is, therefore, easier to compare. However, PSCs operate in a complex environment. Providing security during contingency operations and war involves a lot of uncertainty, where changing conditions on the ground could affect not only the services provided but the very requirements that the government sought to contract.[26] PSCs often have trouble knowing costs and predicting profits. One alternative is the cost-plus contract. In cost-plus contracting, a PSC can pass on operating expenses to the government. In other words, when PSC costs grow, their profits remain unaffected. As Carafano (2008) points out, PSCs might be more "willing to undertake risky contracts where costs might spiral out of control" (77), but there would also fewer incentives to lower expenses.[27] The other alternative is the sole-source contract, where the contract is let to one company without a competition. During a national emergency, the US DOD can let a sole-source contract.[28] The problem with sole-source contracting is that it removes the very market forces proponents of outsourcing rely upon. Fortunately, since 2007, the Iraq security contracts were awarded under competitive bidding.[29]

Nevertheless, even when PSCs are contracted in open bidding, these contracts do not adjust easily to changes in the state's requirements. By their nature, contracts are valuable for delivering a product or service to order. Furthermore, whenever a principal contracts for services to an agent, in this case the PSC, there is always a problem with incomplete information, which compounds the issues of hiring PSCs in contingency operations. Like other outsourcing, the government must establish contractual monitoring and oversight to ensure that even the original contract is being properly fulfilled, and these increase costs. Contracts are also rather inflexible; anticipating contractual changes is difficult and, of course, costly.

Some might argue that PSCs are profit motivated, but so too are other entities that provide products or services. Much as a doctor may make money and be motivated to save her patients, a security company's intrinsic motivation is to make its clients more secure. While it is naïve to think that PSCs are solely motivated by profit, it would be equally absurd to ignore that profit does matter. In the security arena, the business nature of the company can conflict with the PSC's intrinsic motivation to provide security. In other words, the financial nature of the contract itself informs how the PSC fulfills it. Profit directs how operations are resourced and conducted. PSCs, therefore, chose the more profitable solution over the most effective one. Finally, framed by unforeseen contingencies, wartime contracts are unfortunately rather inflexible. A cost-plus alternative seems to counter the very cost savings the original PSC contract was intended to provide. While implementing contractual control would alleviate some of these financial issues, the process is difficult at best, and history has so far has not provided good examples of making it work.

Traction

Another consequence of doing business with PSCs is known as traction, where the profit motivation of the PSC not only can affect the current contract but also encourages that the PSC remain profitable by earning contracts (or renewing the same contract) in the future (Singer 2003, 96). In order to be hired in the first place, the PSC not only needs to demonstrate they can fill the current contract but also recognize that the principal is more willing to invest in a company when the principal can observe that the PSC will be in business for some time to come. Establishing a long term relationship between the government and the PSC means increased profitability for the PSC and future contracts with the government.

However, the anticipation of future contracts can not only build an increased reliance on the PSC, but the same motivation for future contracts can affect how the PSC conducts operations. On the one hand, a PSC might be motivated to do well in order to garner future contracts; while on the other hand, how the PSC fulfills the contract may end up being more of a marketing tool for future business rather than the local effects of the current contract. For example, in Iraq, Blackwater (now Academi[30]) claimed that it has never lost a protected person; however, critics point to the disruption to Iraqi lives from Blackwater operations. The Coalition Provisional Authority (CPA), furthermore, was not only hiring PSCs and attempting to control PSCs in Iraq but also establishing policies requiring other companies and agencies to build private security into their bidding for CPA funding. Before the questions of whether the US should hire PSCs or how they could be controlled could even be answered, CPA policy was encouraging the increased and future hiring of private security.

Not an ideal market

PSCs do not operate in the ideal, free marketplace. While the military operates with controls ranging from specific domestic and international laws, PSCs are primarily regulated by the market. Traditionally, in the PSC industry, as part of the larger military industry, there is no instant supply and demand. The PSC industry is an entrenched industry, where competition is erratic, and the players remain basically the same. While the company names may change, the key players—those bidding on and executing the contracts—are essentially the same, with long term contracts passing between a few large PMCs as the contracts are re-let. As Surowiecki (2004) writes, "Outsourcing works well when there's genuine competition among suppliers." But, the limited players of an entrenched industry limit competition, where competition should be a self-regulating mechanism. Because there are limited players, and therefore limited competition, the self-regulating mechanism of the market is not sufficient for PSCs. The lack of market regulation increases the costs of providing oversight mechanisms, and the lack of competition may result in not receiving the best price for services rendered. Because the market is not ideal, competition enabling self-regulation will not work as PSC proponents advertise. The lack of market regulation increases the costs of providing oversight mechanisms, and the lack of competition may result in not receiving the best price for services rendered.

Defining contractual success

Identifying successful fulfillment of security service contracts is problematic, much more so than for delivery of a product or even for services rendered in other areas. Much like defining success in Iraq, or determining whether the United States is more secure since 9/11, or more relevantly, whether ArmorGroup was sufficiently pro-tecting the US Embassy in Kabul, (POGO 2009; Warner 2009)[31] PSC contract success is difficult to define. How does the principal know when the contract is fulfilled? How should the contacts' measures of performance be measured? Much of the potential issues with defining success may be improved with better contract writing, but often the desired successful results are often more complex than a simple metric. For example, Blackwater merely stating that they have not lost a protected person or that a PSC-guarded base has not been attacked are both due to a number of factors—perhaps unrelated to the security contracts' fulfillment. The problems of the defining contract success combined with the contracts' inflexibility makes adapting to change in security situations much more difficult.

Becoming dependent on PSCs

Continued reliance on PSCs raises another concern. The government may become dependent upon privatized security. This PSC dependency can lead to several problems. First, building upon the problems of contractual uncertainty, traction, and inability of defining contractual success (or failure), increased reliance on PSCs while perhaps providing flexibility in the short term, can limit the government's military options. Second, in terms of principal agent theory, the PSC might gain dominance over the agency that hired it. Third, at the worst possible time, a private company may abandon the client. And, even if a PSC may not willfully abandon its client, a fourth problem may occur as to whether the principal can find a replacement if the PSC fails in its contract.[32] Fifth, once the security function has been outsourced, the government may lose the in-house ability to perform that same function. Private actors can accumulate knowledge and simply transfer it away from the public sphere. Finally, because hiring PSCs is a delegation of force previously performed by the military, the danger of the corresponding transfer of authority with the delegation increases.

Walkouts, strikes, and dropped contracts

PSC dependency also incurs negative short term consequences. As with any business contract, private security contracts also run the risk of walkouts, strikes, and dropped contracts, but the consequences of these potential pitfalls are even greater during wartime and contingency operations. While contractual breaches are not the norm, they have occurred regularly enough to warrant some investigation. For example, in 2004 60 South Korean subcontractors north of Baghdad quit work on an Iraqi electrical project when two fellow workers were killed (Surowiecki 2004). British Royal Navy shipyard repair reportedly had to deal with threatened labor walkouts (Singer 2003, 162). In addition, in May 2006, 450 British PSC Control Risk employees, who provided personal protection in Iraq, threatened resignation (Isenberg 2009, 86). Gurkha guards in Kabul have threatened to quit guarding the US Embassy in Kabul, and apparently "nearly 90% of the incumbent US/Expats left [the ArmorGroup contract in Afghanistan] within the first six months of contract performance" (POGO 2009).

And if the demand is great enough, they may be hired by someone else. Because the PSC is a business, further attention needs to be focused on how mergers, bankruptcies, foreign acquisitions, and other business decisions affect the security service contract. Because of the immediate negative effects of a walk out or broken contract, these risks must be evaluated prior to the contracts being let.

Overcharging and monies wasted

Along the same lines as contract breaking is potential problem of financial abuse. Much of the PMC media debate focuses on issues of monies wasted, and with proper oversight, the risks of these financial abuses could be lessened.[33] Nevertheless, poor contract management leads to wasteful spending of billions of dollars and can divert limited resources away from other important US counterinsurgency efforts (Walker 2007). Schwartz's 2009 CRS report on Defense Department contractors states that "wasting resources that could otherwise have been spent on social services and economic development may limit the effectiveness of US efforts," and "may also result in increased fraud, which could similarly undermine the credibility of the US in the eyes of the local population" (12). A PSC proponent might counter that the government also wastes money, and fraud and overspending are some of the reasons for the privatization push. I concur; however, it is worth mentioning that both overcharging and wasted spending to show that the hiring of PSCs is no guarantee that costs will be saved. Consider the report that guarding the US Embassy in Iraq costs six times what guarding it by military forces would cost

("Corporate Warfare" 2009). Without true cost savings, the PSC argument loses some traction.

Boom or bust cycle of war and peace[34]

As we watch the stock market for the bull and bear cycles, a state is often shifting through periods of war and peace. Often the market fluctuations correspond to the war-peace cycle; often they do not. A price rise during a "boom" period could cause excessive burden on the state. This would be especially costly if the PSC drops the contract (as I mentioned earlier) or even if the PSC decided to renegotiate the contract in a time of crisis. In the middle of a conflict, the government as the client is in no position to counter because of the increased risks of being able to immediately re-let the contract or cover the service in-house. (The renewal of ArmorGroup's Kabul contract may reflect this phenomenon.) While different PSCs' leadership may disavow using this kind of business tactic, it is a common practice in the business world, and the consequences of employing this business practice in the security sphere would be keenly felt.

Control and security privatization

In the previous section, the focus of the discussion of the negative consequences of privatizing force through hiring of PSCs was cost-based. However, as the reader has probably discerned, many of the financial cost issues mentioned seem closely intertwined with issues of oversight and government control of the PSC contract. I now turn to the issues of control and security privatization in this next section, focusing on the problems that arise in the practice of subcontracting, the introduction of multiple and transnational stakeholders into the force commodification equation, the control complications that arise from hiring third country nationals, as well as the control of individual PSC contractors who often have shifting alliances and are often treated as independent.

Sub-sub-contracting

While much attention is focused on the larger PSC companies, such as Academi and ArmorGroup/G4S, what some fail to realize is that once the contract is let, many of the PSCs further subcontract for security services. For example, the ArmorGroup contract to guard the embassy in Kabul contains a subcontract for Ghurka guards. In 2004, of the 60 "PMCs that the CPA identified as working in Iraq, only eight had direct contracts with the CPA" (Isenberg 2009, 26).

The common practice of subcontracting is a sound one in theory. Properly used, subcontracting allows for greater efficiency and is used in a wide variety of business areas, from manufacturing to construction to software development. Nevertheless, in the security service industry, subcontracting increases the challenges of control, accountability, and oversight, working "two, three, four, or even more levels down the contracting chain" (29). In the 2004 Blackwater Fallujah incident, identifying whom precisely the contractors were providing security for was complicated by just this effect. Although stipulations limiting subcontracting can be incorporated into the PSC service contract, this common business practice is hard to shake. Using military forces has the advantage of having a clear chain of command providing lines of responsibility to the lowest level. This problematic phenomenon was recognized in Iraq, where the CPA attempted to rectify the problem. Its solution was hiring yet another PSC, Aegis, to coordinate all movements of PSCs.

Multiple (and transnational) stakeholders

Control issues are further exacerbated because the PSCs operate globally. The introduction of multiple and transnational stakeholders introduces another set of concerns. States are not the only principals that use PSCs. Often states contract PSCs for military services to international organizations or other states on their behalf. Doing so creates multiple often competing principals. Certainly, the United States hires its own PSCs, but they also hire foreign ones, and PSCs are hired by foreign governments and even other PSCs. When the PSC contract is let in Washington, for example, but executed in Pakistan, the principal paying for the contract does not directly see the service being provided. If a PSC is publicly held, then its shareholders come from a wide variety of backgrounds and are not necessarily beholden to the government who initiated the contract. Furthermore, PSC shareholders may be external to the conflict; they may have no interest in resolving the conflict other than ensuring financial profitability of the PSC. While the contract itself may force PSC compliance, the differing interests of multiple principals increase at the risks of divergent influence and control over the PSC execution of the contract. Therefore, multiple principals may increase issues of control.

Transnational markets in general usually create challenges to oversight and control, and the PSC industry is no different. This raises two further concerns. First, a state may control a particular PSC during a particular contract, but there is no guarantee that this control extends to the PSC's other interests, nor to other PSCs. Second, in cases where a weak state hires a PSC, or another stronger state hires one on its behalf, multiple principals operating internationally further diffuses control (Avant 2005, 75; 130–31).

While proponents of PSCs, especially in the US, argue for the loyalty of the US-based PSCs, we would be prudent to remember that these companies are transnational, and there are multiple principals involved. Erik Prince may have argued that Blackwater will only work for United States interests, yet "contractors of the late Middle Ages had obligations to the feudal lords even as they sold their services elsewhere. By necessity each of them had two masters: his feudal lord and the one with whom he made a contract" (246).

Third country nationals

Not only do problems arise from PSCs being transnational but there are also additional complications arising from their global hiring. Contractors from country

A, who are hired by country B, and work in another country C are commonly called third country nationals (TCNs). Many of the PSC contractors in Iraq and Afghanistan were and are TCNs.[35]

The numbers alone are not the issue. First, TCNs are citizens of other states, often ones not parties to the conflict,[36] and are potentially problematic under international law. Second, TCNs incur the same difficulties with control as the subcontracting practice. Third, TCNs are bound only by the subcontract to perform services; for example, the Gurkha guards in Afghanistan are hired by ArmorGroup, not by the original principal, the US State Department. Because the TCNs are subcontracted, hiring of TCNs raises a further area for friction: screening and selection of personnel is left to the PSC. Furthermore, large-scale use of TCNs lends credibility to the appearance the PSCs are, or at least are hiring, mercenaries.[37]

"Independent contractors"

Unfortunately, the problems associated with contractual control do not end with the hiring of TCNs. Some PSCs themselves are adding to the issue as they attempt to distance themselves as a corporation from illegal acts performed by their employees. In 2008, Blackwater reported to Congress that its security guards in places such as Iraq are "independent contractors." Remarkably, as Rep. Waxman (D-CA) reported:

> Blackwater has claimed in official communications that its security guards are 'in no way directly supervised or controlled by Blackwater'; that they 'do not report to any of the Blackwater entities regarding their work in the field'; and that they 'do not report to Blackwater regarding their operations in country.'
>
> (US Congress 2008a)

Therefore, if this testimony accurately depicts the business relationship, then we are left with the absurd conclusion that Blackwater's only real responsibility was to pay the independent contractors.

In other words, a PSC such as Blackwater, seems to be claiming that they function as a hiring service when fulfilling these security contracts. Control, or responsibility for lapses of control, then falls back upon the principal, in this case the US State Department. These claims seem to imply that in 2009 there were 10,422 independent security contractors operating in Iraq.[38]

Shifting companies

PSCs, like other businesses, transform themselves. While contractual controls may be put into the contract to ensure that the contracted service will be provided when the company is acquired by another, other problems remain. If a PSC, C, violates a contract and is banned from bidding on further government contracts, C can dissolve and C's employees can go work for another PSC, C2, or even form a different PSC, C3, to bid on new government contracts. This allegedly occurred with Custer Battles in Iraq (Isenberg 2009, 90), and some theorize that this is the rationale behind Blackwater's transformation into Xe then into Academi.

Since PSCs also are transnational, they often establish subsidiaries in many other countries. (For example, British PSC DSL had approximately twenty of these subsidiaries.) PSCs desiring to avoid troublesome state regulation can break up and move somewhere else. Finally, once an employee leaves one PSC, she can gain employment in another, much like Daniel Fitzsimons, who was fired from both Olive Group and Aegis, and later joined ArmorGroup in Iraq, where he subsequently killed two coworkers (Gillard 2009).[39]

Who guards the guards?

The problems of shifting companies and of controlling independent contractors raise what I term the problem of guarding the guards. In contingency operations or war, proper oversight becomes not only more important but it also becomes vital to prevent fratricide and other tactical mistakes as well as ensuring compliance with the commander's operational campaign plan. With any outsourcing, additional layers of control are required by the principal to ensure that the contract is meeting the requirements in a manner that the principal desires. Repeatedly, however, even this oversight function is outsourced because the principal lacks the resources or expertise to provide it themselves.

When the Government Accounting Office (GAO) reviewed the Army Contracting Agency's Contracting Center of Excellence, which oversees procurement, it identified that 42 percent of the Army's Contracting Center procurement specialists are themselves contractors (Isenberg 2009, 89). While the GAO reports that relying on contractors creates the risk of loss of government control and accountability, the GAO itself hires contractors to monitor other contractors. "Moreover," Verkuil (2007) writes, "in the military setting, the contracts themselves are often secret or severely circumscribed in disclosure, making it difficult for the public to monitor" (149).

The problem with making force a commodity through hiring PSCs is that it also makes oversight a commodity. Verkuil argues that government oversight "as a public value has been diminished, if not eliminated. Outside monitors may do acceptable work, but they must themselves be monitored" (149). Because privatization of force redistributes the control of violence, the use to private guardians changes the question of who guards the guards.

Private security and the changing international landscape

In the last sections, I have explored some of the problems and consequences of hiring PSCs in terms of financial costs and multiplying control problems. I now turn to the potential that outsourcing force to PSCs may allow governments to conduct foreign policy by proxy—a view that proponents of PSCs, nevertheless, believe to be absurd. Then, I focus on the impacts of privatized security in the changing international landscape, specifically in the developing world. Next, I will consider how widespread hiring of PSCs actually increases the threat of violence from private groups, and I will discuss the effects of PSCs in what Mary Kaldor (2006) calls "new wars." I conclude this section with a discussion of what happens when the primacy of security encounters a competitive market.

Foreign policy by proxy

In the US, the hiring of a PSC is a market transaction by the executive branch. Contracts less than $50 million are not evaluated by Congress. Because PSCs are armed and can use force, especially in the conduct of foreign policy, combined with the inherent control problems, their employ can allow for executive agents to bypass normal government checks and balances.[40] For example, when Congress establishes a troop cap for a country, A, limiting the number of military personnel allowed to be in A, hired PMCs can and have been used, essentially going around the congressional limitations. When President Obama called for a surge of 30,000 soldiers for Afghanistan, he did not mention the estimated 58,000 contractors required to support the troops.

Proponents would counter that the Freedom of Information Act and other legal oversight mechanisms, as well as a motivated and open media, prevent foreign policy by proxy. Perhaps this is true—at least in the US. But, incidents such as the Iran Contra Affair suggest that this may not always be the case. While some may point to the need for and examples of congressional approval even for operations short of a declared war, such as the Iraq invasion, the US has not declared war since World War II. US supporters of PSCs should also remember that the US and her allies are not the only international entities capable of and resourced to hire PSCs. In other words, can one expect a similar amount of transparency in say, Myanmar, Iran, Syria, or North Korea, or in wealthy corporations or criminal organizations?[41]

International reputation plays an important role in international relations. Furthermore, while the idea of foreign policy by proxy sounds like a good spy novel, the capacity is there should the conditions be right. Western democracies themselves may be victims of it; more probable, however, is foreign policy by proxy within other states or by non-state actors. One might ask whether Syria is using the Lebanese Hezbollah as a proxy force—a non-state armed force. Tyler Cowen (2007) in the *New York Times* even suggested, "When it comes to Iraq, we have yet to see the evidence of large practical gain in return; instead, use of contractors may have helped make an ill-advised venture possible."

Security for those who can afford it

Early in his book, *Private Sector, Public Wars*, Carafano (2008) suggests that the "use of contractors in combat can become one of [the US's] greatest competitive advantages in the twenty-first century" (12). Moreover, if the hiring of PSCs results in greater benefits than the costs under inspection in this project, then PSCs may be a viable alternative. If Carafano's assumption is correct, then PSCs may serve well for wealthy countries like the US. However, this raises another important question.

When force becomes a commodity, something that is bought or sold, it requires consumers. While some consumers can afford to purchase the commodity, others are less fortunate. This phenomenon may lead to a stage where there are those who can afford security and those who cannot. While having the means to hire PSCs may bode well for Carafano's vision of US foreign policy advantage, it has quite different results in the developing world.

In weak or developing states, there tends to be a number of internal and external principals that participate in the security sphere. In the developing state, transnational corporations, NGOs, other states' foreign military assistance personnel, PSCs, paramilitary groups, and the state's own security forces all have a hand in the security situation of the state. Externally, the different players introduce the same multiple principal concerns discussed earlier. In terms of transnational corporations' dealings in developing states, these corporations tend to hire external PSCs or local security forces to protect company personnel and key installations. External control of force combined with considerable financial influence might generate loyalties to those external to the local government. Some PSCs proudly declare that they only work for legitimate state governments. However, if PSCs are hired by the local government, then the PSCs may be enabling the status quo. While not always problematic, the use of PSCs may preclude legitimate resistance movements from taking hold; worse, the PSC may be used as a tool for internal repression.

Transnational corporations and NGOs also generate other problems internally. While often inadvertent, the external money they bring to provide local security is often funneled to the most capable or well-connected in a certain area. Much like the warlords in Afghanistan or the tribal leaders in Iraq, the influx of external monies can create privatized security enclaves that are separate from and in lieu of the public security supposedly provided by the state.[42] Consider the Syrian conflict, where the EU and the US are offering military aide to the rebel forces. Two further problems arise. First, security threats are now deflected into poorer areas. Second, the chance that developing states create public security institutions decreases, as parallel, and often competing, security is provided by private means. Or, as is common practice, the state must choose between the financing of public bureaucracies and privatized security, especially if internal threats loom.

Ironically, even in strong state-sponsored development, where the stronger state is coordinating and executing another's development, such as the United States has attempted in Iraq and Afghanistan, the wealthier state's hiring and subsequent use of PSCs outside the local government's control, "prevents the citizens of post-conflict states from controlling the armed agents working in their territory" (Percy 2007b, 237). The new Libyan government seemed to fear this very problem, refusing to allow armed contractors in large numbers in post-Gaddafi Libya (Risen 2012). Therefore, almost incredulously, as Kevin O'Brien (2007) writes, private security is "reconstructing the state and its provision of public security to its citizenry while, at the same time, deconstructing the state's monopoly on armed force and public violence" (33). While the lack of security provides the reason for hiring PSCs initially, by hiring PSCs, the external principals themselves undermine the very role of public security the developing state requires.

Threats from private groups

The struggle in Iraq to eliminate private militias, the international search for a solution to the pirates off of the Somali coast,[43] and the Colombian and Mexican governments' respective conflicts with drug cartels demonstrates how states and international organizations are expending large amounts of energy and resources to eliminate competing private forces. Nevertheless, the hiring of PSCs is reintroducing private force back into the equation. While many supposed legitimate PSCs may not provide an increased threat by themselves, past experiences with other PMCs has shown that, depending on their employer, they might pose new threats, and in the future, some PSCs may themselves be large enough to compete with states, including former employers.

In her book, *Making a Killing*, Drohan (2004) provides several examples of how transnational corporations have used force and how several have been linked to armed conflict. Union Minière in Katanga, Lonrho in Mozambique, and Shell in Nigeria are just three of the examples that she explores in detail. As Singer (2003) suggests, PSCs "represent the next logical step on the services side, and, as such, become simply a more direct extension of the power of outside corporations" (189). Furthermore, PSCs themselves chose their employers. Some work solely for their home states, conducting security operations in full accordance with their home governments. Others have not. Some have worked for other, non-state organizations and companies that have been involved in drugs, arms, insurgencies, and other illicit projects. Without strong legal regulation and with the right money offered, PSCs will be tempted to continue working for other, more dangerous groups.

Much like the mercantile companies of the past, PSCs have the potential to grow in strength, and, as EO did in Sierra Leone, to become stronger than the state that employed it or its neighbors. While there is of course no guarantee that a PSC would directly compete with or overthrow its employer, it is very certain that the interests of some PSCs will conflict with some states' interests. Armed with their own PSCs, corporations, non-state groups, or even individuals with the financial means could again be international players. It also means that it is possible that large PSCs could in effect function as their own international force themselves, not under the employ of a state.

New wars and PSCs

In her book, *New and Old Wars*, Mary Kaldor (2006) describes the term "new wars" to explicate the wars that are not fought between states for traditional reasons such as gaining territory. Rather, new wars reflect other motivations: identity politics, or power over or exclusion of another group. In new wars, whole war economies are often created; some new wars are fought purely based upon economic calculations. Growing numbers of PSCs are moving into this new war dynamic. Because the new wars are fought on a smaller scale, combined with a lack of stronger states or international organization means or interest, the PSCs themselves are altering the landscape.

Economics of course has always been intertwined with war and security, but now there is a nexus of those with the means seeking to purchase force as well as profit motivated security companies willing to sell their services. Military power is becoming increasingly fungible. New wars are generally more ideologically motivated; they involve criminal activities, and unfortunately they are often marked by a large number of human rights violations.[44] While PSCs could provide stability in a certain region for a time (e.g., peacekeeping operations by PSCs),[45] the cycle will increase demand for PSCs, and generate more competing private security companies, therefore increasing opportunities for other, less savory, purchases of force.

Public good vs. the private (profit) good of PSCs

Delegations of government authority to private hands are not simple transfer decisions. They are decisions that potentially transfer sovereignty. They should come with strings attached to ensure fairness at the individual level and accountability at the political level. The debate about public versus private, therefore, becomes a meditation on our constitutional values.

(Verkuil 2007, 81)

Cost and profit motivation are not the only things that draw a distinction between a private contractor and a civil servant. Private contractors are endstate focused. Certainly, they deliver a product or service in exchange for money, but the focus is on the ends of contract fulfillment. A civil servant, alternatively, agrees to accept instructions—in this case in exchange for a wage not limited by a specific contract. Determining whether private contractors or civil servants will be most effective is based upon the nature of the task/service. "Contracting makes sense," Avant (2005, 48) notes, "if the government is exactly what it wants and cares more about the ends than the means." On the other hand, a civil servant is preferable "if the government cares about the means and wants its agents to follow a set of guidelines for how to go about providing a service."

In other words, if one is concerned with *how* a contract will be filled, she must outline all the specifics, including these considerations in addition to measures of performance in the contract itself. Not only is this process time consuming, perhaps resulting in an unwieldy contract; but it may also negate the increase in efficiency the principal sought with outsourcing. Further contractual changes to account for unforeseen second order effects could also turn costly. Agents of the government may perform the same service, but their scope of work can be quickly adjusted without contract negotiation. Which is the better option depends upon the desired requirements.

For example, to a state that hires a PSC for security, security is not merely an important task, it is a necessary one. The hiring principal must balance the primacy of security with the very business competitiveness that caused the government to look towards outsourcing in the first place. Often these cannot be balanced. Furthermore, cost savings alone are not the only consideration. Neither are concerns of control. The decision to privatize force often blurs the line between the public good and the private good as contractors become essentially agents of the state. To illustrate some of the consequences of these conflicts, in this section I will discuss the hiring PSC effects upon a state's alliances and security assistance programs, its conduct of counterinsurgencies, and a state's international reputation. In particular, I focus on when the private sphere and the public sphere conflict with one another.

Alliances and security assistance

Even long-standing international alliances are often fragile. Within the alliance itself there are concerns of power-sharing, financing, treaty obligations, and nuances of diplomacy especially over decisions to commit the forces of the alliance. Internationally, alliances allow smaller states to retain some authority against stronger states and other alliances. Historically, alliances have sometimes resulted in wide-scale conflicts, such as the First World War, but have more often maintained a kind of balance of power in certain regions. Adding PSCs changes the equation; the question is to what degree. No studies have been conducted on the impact of hiring PSCs on these alliances. However, one possible result is that states may alter the balance of power within alliances or between alliances through the hiring of private forces. Burden sharing within alliances might become less necessary as states that can afford private force hire PSCs or the alliances themselves hire them. Recall that the prime minister of Papua New Guinea turned to Sandline because its traditional ally, Australia, declined their request of support for specialty military training and equipment (Spicer 1999).

Another issue is that alliances and coalitions rely upon common training, processes, and shared experience to be militarily effective. The use of PSCs as proxy security assistant providers places these relationships at risk. Much of the impact of PSCs on alliances also affects a state's reputation but manifests itself when alliances or coalitions conduct complex operations like counterinsurgencies.

PSCs in counterinsurgencies

Allegations of misbehavior by ArmorGroup employees in Kabul, combined with doubts that they were adequately fulfilling their embassy security contract, highlight potential problems with the employment of PSCs in counterinsurgency campaigns like those in Iraq and Afghanistan (POGO 2009). Winning the hearts and minds of the local population, thereby gaining legitimacy in their eyes, is a fundamental part of the operation of waging a counterinsurgency. The US Army's Counterinsurgency Field Manual further articulates, "Security force abuses and the social upheaval caused by collateral damage can be major escalating factors for insurgencies" (2006, 1–9).

Unfortunately, as demonstrated by the 2007 Blackwater shooting in Baghdad that left 17 dead (Glanz and Rubin 2007), this best practice is often in jeopardy. Furthermore, as the CRS "Contractors in Iraq and Afghanistan" report points out, "abuses and crimes committed by armed private security contractors and interrogators against local nationals may have undermined U.S. efforts in Iraq and Afghanistan" (Schwartz 2009, 11).

A PSC proponent would point to similar episodes of mistreatment of the local population by the military to counter that PSCs are not alone in their counter-productive activities. Often, PSCs—and PMCs in general—are more professional (and often more capable) in their daily duties than a state's own security forces; many of them served honorably in their state's military. Nevertheless, hiring of PSCs raises additional worries. First, the local population often draws no distinction between PSC contractors and the military. Any mistreatment by either can be attributed to the "occupation."[46] Second, how PSCs provide their services, their methods, have broad effects. The use of excessive force or the threat of force during the execution of the contract may, in counterinsurgencies especially, erase the hard won goodwill that the military and government are trying to achieve. While some may argue that better controls through improved contracts and oversight will eliminate this prob-lem, (1) this has not been the case yet, and (2) I argue that the military itself is better suited to adjust to an evolving situation. In the case of counterinsurgencies, being able to adjust how one accomplishes the mission to meet the overall commander's intent for the campaign is essential; for a PSC, this would require a change in contract as well as a change in how the PSC conducts business.

International reputation

Allegations of abuse at Abu Ghraib created a lot of controversy, generating ill will internationally. Representative Jan Schakowsky (D-IL), in a letter to President Bush in reference to Abu Ghraib, wrote: "I maintain that the use of private military contractors by the United States as a misguided policy that costs the American people untold amounts in terms of dollars, US lives and is damaging our reputation with the international community" (quoted in Isenberg 2009, 114).

Much of the problem in terms of reputation stems from the historical ties of PSCs to mercenaries. As Isenberg (2009) points out,

> With the exception of a few companies that have used ex-British Army Gurkhas … the recent trend is to strip Third World armies of full battalions in order to be the lowest bidder. It lends a bit of truth to the accusations that [MNF-I] is paying foreign mercenaries.
>
> (165)

In addition, the fact that there has been misconduct, as well as issues of over-billing and other contractual oversight problems, seems to confirm the perception of many that PSCs "are wholly independent from any constraints built into the nation-state system"; Percy (2007b, 58) quoting Zarate (1998) continues, "The element of accountability is the tacit standard that underlies the international antipathy for mercenary activity and truly determines mercenary status."

States often cannot separate themselves from the conduct of PSCs when they have hired them in the past. Furthermore, the conduct of the PSC often reinforces the mercenary reputation of the company and the state that hired it. Alternatively, the use of PSCs may increase a state's ability to act unilaterally, without the prior investment onto a larger standing military force. This ability may affect, for example, the Obama administration's attempt to assure the international community that the US has no desire to "go it alone" in current or future conflicts. The perception of another state's intentions and its international reputation have always played a significant role in interstate diplomacy, and the use of PSCs further affects both perception and reputation.

Conflict between public and private spheres

Some have proposed that liberal democracy is actually about answering the question of what can be made public and what should remain private. The notion of public and private spheres is enshrined in the US Constitution and also grounds much of the common law. Often, the lines are clearly drawn; other times the spheres seem to collide. With a return to the use of privatized force, however, the public-private debate has raised some new tensions in domestic and international politics and norms. Consider the cleanup after Hurricane Katrina. When Blackwater patrolled the streets of New Orleans after Katrina (Sizemore 2005), were its agents private or public? Can contractors enforce the law? In the United States, the Posse Comitatus Act limits the use of military forces for domestic law enforcement. Does the use of PSCs domestically challenge Posse Comitatus?

Furthermore, providing for the common defense is an accepted function of government. Security becomes a public service, and it requires a "special public trust." The public perception is that security is a public good, one to be provided by the government, not reliant upon the market. Consider the Bush administration's decision not to outsource airport security; instead the Transportation Security Administration (TSA) is a government agency. Because of the perception that security is a public good and that providing airport security is a symbol of the government's commitment to public security, it was not privatized (Verkuil 2007).[47]

While it seems that one can argue that as long as the state does provide for the common defense, it should not matter whether the forces are public or privatized. Yet, one of the apparent concerns is that when the state outsources to PSCs, it is delegating an essential responsibility. Even if it is an actual delegation, it can be perceived by its citizens as an abdication of responsibility. This issue is compounded because the public good is not identical to the private good. Recall that the organizing intent of PSCs is to generate profits, as any business does. It does so by providing a security service. But, then it seems as though the citizens of a state no longer "enjoy security by right of their membership in a state. Rather, it results from the coincidence between the firm's contract parameters, its profitability, and the specific contracting members' interests" (Singer 2003, 226). Privatized security may thus challenge the often presumed republican ideals of the rights and responsibilities of the relationship between states and peoples.

Some objections

Before I conclude, I will now turn to some anticipated objections to this view, including that Weberian states no longer exist, that there is no monopoly of force, that force is already a commodity and PSCs are a mere evolution of this trend, and that licensing or heavy regulation of PSCs will eliminate the problems that I have identified. I will also explore whether security alone should not be privatized or should the prohibition apply to other military functions as well. Finally, I will address the argument that the market can moderate the behavior of PSCs.

There is no state (or at least a Weberian state)

As I discussed, my argument relies upon the assumption that states have a monopoly on the legitimate use of force as well as the assumption of state sovereignty. An objector might posit the extreme view that assuming Weber was correct and that hiring a PSC somehow erodes a state's monopoly of force, then states—at least states as Weber conceived of them—no longer exist. I of course disagree. Walzer (2008) writes,

> If we want to maintain accountability in war, then, we had best take a statist view of military activity .... The state is the only reliable agent of public responsibility that we have. Of course, it often isn't reliable .... Still, there isn't any agency other than the state in the contemporary world that can authorize and then control the use of force—and whose officials are (sometimes) accountable to the rest of us.

Furthermore, I agree with Walter's proposal above that (1) states do exist, (2) there are currently no better options for authorize use of force than the state, and (3) barring any significant change in the international landscape, such as a truly empowered UN, that the state retains the monopoly of force and should continue to do so.

Some state proponents point to recent game theoretical work to support the case for states. Avant (2005) writes that these studies reinforce

> The necessity of the state, and its control of violence, for productive societies. They argue that societies cannot be both peaceful and prosperous without a state. Stateless societies can be peaceful, but only if they are poor. As economic prosperity grows, so do the incentives for raiding and predation.
>
> (46, footnote 18)

However, arguing for states themselves is beyond the scope of this discussion. The question at hand is whether the hiring of PSCs eliminates the monopoly of force, and if it is this monopoly itself that defines what it is to be a state, then states are in danger of extinction.

While this may not be the case today, it is conceivable that widespread hiring of PSCs could dramatically alter how we perceive states and their functions. The problem then is not that states no longer exist; rather, the problem is that long term reliance upon privatized force may change conceptions of statehood in the future. While I do not view this as an objection, it does indicate another consequence of commodifying force that those considering hiring PSCs should keep in mind.

There is no monopoly of force

This objection argues that this monopoly of force is an illusion, or if real, is not a monopoly at all. Consider organized crime, terrorist groups, insurgents, and other non-state actors. States may have had a monopoly on force in the past, but the wars of the twenty-first century highlight that the nature of belligerents has changed (Bailes, Schneckener, and Wulf 2006). I must concur that the state's monopoly of force is diminishing by these same actors. However, I agree with Walzer that while perhaps anachronistic, states continue to be the global political framework that we have. Furthermore, the monopoly of force is the *raison d'être* of states. This monopoly of force, while not perfect, keeps the jus ad bellum decision to resort to war a "public responsibility." (I will return to this problem in Chapter 6.)[48]

Nevertheless, considering other governmental functions that the state has a monopoly over, such as currency printing,[49] we could ask do states have a monopoly on the printing of their currency? While it may seem strange to say that a state has a monopoly on printing its currency, it is clearly a government function that as a whole should not be privatized. What if the state outsources the printing of currency to India? Why would this outsourcing be okay, yet outsourcing force not be? Consider another example. Let us assume that the state has a monopoly on the delivery of mail. Yet, the post office can and often does contract for delivery of mail in especially rural or mountainous areas. Should we say that there is no monopoly?

While these are perhaps good questions, there are two problems with this line of reasoning. First, as I pointed to earlier, there is a separation between the outsourcing of the service itself and the outsourcing of that service's management and oversight. In the cases of currency printing, postal delivery, or even when the IRS uses contractors to collect delinquent taxes,[50] the contractual oversight responsibility still resides with the appropriate government agency. When this oversight is outsourced as well, the multiple principal problem and the issue of who guards the guards is introduced. Likewise, assuming that the benefits outweigh the costs, where the costs include mechanisms for adequate oversight, states can hire PSCs without losing their monopoly of force. That states have thus far been unsuccessful in doing so in terms of oversight indicates that hiring of PSCs is not only not beneficial but also negatively affects the state's monopoly of force.

Second, privatizing force is different from the other more innocuous outsourcing examples. These other kinds of outsourcing have become widely accepted. However, privatizing force, such as by hiring PSCs, risks putting security outside of public control.[51] As noted in the last section, the risks of the negative consequences of hiring PSCs are far greater than those caused by a failure of mail delivery or tax collection. There is simply more at stake.

Force already a commodity: PSCs are an evolution in forces

An objector might persist, arguing that since states or local communities employ police and security guards that this forms a delegation of the monopoly of force. Furthermore, they might contend that the hiring of PSCs is merely a delegation of this monopoly of force into a wider arena.

Those holding this view, however, naively equate domestic security and use of force with a state's security and use of force abroad. She would be correct to view, for example, police as an indication of delegation, but (1) it is not a total delegation without continued responsibility, nor (2) is this domestic delegation the same as a delegation of force beyond the state's borders. She might counter that many states, especially criminal or illegitimate states, blur the responsibilities of domestic and interstate use of force, and I concur. However, democratic states have a history of separating the ends, functions, and areas of responsibility between domestic and interstate security; that is, the Posse Comitatus Act of 1878.

More telling are the differences in the expected means between domestic law enforcement and state security. Tony Pfaff (2000) draws upon this distinction in his "Peacekeeping and the Just War Tradition." He describes that interstate or military forces operate from the constraint of the maximum allowable force; while police are constrained by the minimum force necessary. For the military, "it is always in the commander's interest to place as much force as is morally and legally permissible on any particular objective in order to preserve soldiers' lives. This means . . . always asking themselves how much force is allowable, not how little is possible."

Alternatively,

> For police the application of force is oriented towards the least amount possible. When police apply force against a suspected perpetrator they are not permitted to use deadly force as a first resort and never if it is the case that the perpetrator is not likely to harm anyone . . .
>
> (2000)

We are of course primarily interested in the employment of PSCs externally, and in this role, I argue that states should not normally delegate the use of force. Interestingly though, one could point to the employment of PSCs domestically, like Blackwater post-Katrina, to show that my police-military distinction is even blurred in the United States. While I agree that we should be alarmed by the employment of armed contractors domestically, and it shows a new problem with states delegating force, I think that this point reinforces my position that states should not delegate their monopoly of force.

Bounty hunters, search posses, deputies, and the black box analogy

Let us look at the state's monopoly of force another way. States, with the need for force, can buy it, delegate it, raise its own forces, call for volunteers, even (like in the old Western movies) deputize. In other words, think of a "black box," where the box can represent any number of types of forces. The state is the input, and the output is the force that the state requires. Call this the Black Box Analogy.[52]

Recalling the proponent's argument for PSCs, the black box could represent

1　regular/standing military;
2　conscripted military;

3   non-conventional forces (e.g., impressed sailors, child soldiers, etc.);
4   hired/contracted forces; or
5   a combination of the above.

Furthermore, the analogy is useful to highlight the differing perspectives on the control force. For example, the box may be black because of the nefarious and secretive role of proxy forces. Or, the black box could represent "adequate oversight" provided by the state. Even this way, one might ask, what difference does it make to the state as to what kind of force a black box represents? Why then are some sources of force considered legitimate while others are not? As I discussed previously, as long as the benefits outweigh the costs, hiring PSCs may be legitimate. The prohibition against their employment is not absolute.

But, the black box analogy is also helpful in bringing to light other issues. First, look at other kinds of security forces. Consider bounty hunters, search posses, and deputies. How are these different, especially when the output is the same? Bounty hunters are also financially motivated. They are the Western version of the mercenary, and like mercenaries, I think most would consider that they are performing a task the government could better provide. Assuming that search posses are motivated by their patriotic duty and not from revenge, they too are performing a government function, but without pay. Because my argument includes non-financial costs and benefits, then in certain cases the use of these posses may be appropriate. Deputized citizens are different from bounty hunters and search posses because they are agents of the government. In a sense, they are licensed to perform a government function. Furthermore, the same consequential argument would note that there may be times when deputizing is a good alternative. One problem with these three kinds of forces, however, is that there is no strong enforcement mechanism to keep someone from walking off the job. Like deputies, some propose that we license PSCs.

How might the licensing, as a limited authority to act on one's behalf, of PSCs change the argument? What if we deputize the PSCs? While many on both sides of the PSC debate argue for better regulation, and certainly a licensing regime could increase the contractual control and oversight of employing PSCs, the argument itself does not change. Currently, an export license issued by the State Department's Office of Defense Transitions Assistance is sufficient for PSCs that are based in the United States; these are not hard to come by. This example highlights the weak end of a licensing spectrum. At the other end, one might suggest making the PSCs part of the military. This option seems to solve many of the control issues, but also removes the services that the PSCs provide from the market that was not only supposed to regulate the PSCs, but more importantly, provide the cost savings and increased efficiency that caused the state to look towards outsourcing in the beginning. A licensing regime would need to fall somewhere in the middle, and its effects would have to be examined to determine if the benefits would outweigh the costs. Perhaps sufficient regulation would make the use of PSCs beneficial enough to remove the all things considered prohibition. So far, however, these licensing mechanisms are not in place.

The market can moderate behavior of PSCs

Much as free-market proponents argue that the market itself can regulate businesses, some PSC proponents, such as Carafano, argue that the market can moderate the behavior of PSCs. Over regulation by the government could also erase the benefits of outsourcing. One might quickly counter that, like the problems with Lehman Brothers, often the market alone is not sufficient to regulate PSCs. However, this dismissal is too hasty. To capitalize on the outsourcing of security, the market, through its supply and demand forces, can provide some regulation. States as the buyers of PSC services do influence PSC behavior based upon consumer demand. Likewise, as part of the security market, states could demand that the PSCs they hire follow international law, including concern for human rights.

Unfortunately, there are three major concerns with this optimistic view. First, at least in the private security industry, there is no clear example that this kind of market regulation is occurring. Let us, however, assume that it could; there would still be the second problem that state-hired PSCs, which practice human rights in their business models and only work for legitimate clients, would leave part of the demand for security unfilled. Market forces suggest that some other PSC, perhaps one that does not proscribe to closely following international law, would fill that demand. PSCs then must consider the trade-offs between how they provide their services. Other companies are available to take their place, such as when Sierra Leone ended their contract with Gurkha Security Guards and turned to EO because they were willing to participate in more than mere training operations.[53] Third, states that hire PSCs, such as the United States or Britain, cannot then condemn another state for doing the same. States will use PSCs in varying ways, just as they use their own military forces to meet their own needs. This seems especially true when a state wishes to discharge a service where deploying their military may be politically sensitive or untenable. Since the market for private force is also open to non-state clients, states may begin to lose their ability to justify preventing non-state uses of private force. By turning to PSCs, states also (perhaps unintentionally) send the message that states no longer exercise sole authority over the use of force.

Benefits and costs[54]

Proponents of PSCs and outsourcing, in general, argue in terms of cost-benefit analysis; hiring PSCs reduces government costs and increases efficiency. Certainly, at least in the short run for specific services, PSCs can make good on their savings. Nevertheless, as I have shown in this section, the benefits and costs associated with the question of privatizing security are not restricted to the financial arena. Issues as diverse as contractual control and international reputation help illustrate the other costs of doing business with PSCs. Furthermore, the financial cost savings attributed to using PSCs are also not apparent. When deciding whether to privatize airport security, for example, "the notion of private contractors conducting safety inspections struck both legislators and the public as a distortion of government responsibilities. When it equated airport security officials with custom officials, Congress in effect endorsed the necessity for public service" (Verkuil 2007, 60). Reassuring the public of the government's ability to provide the security of its citizens was more beneficial than any foreseeable financial saving to be gained by the hiring of PSCs.

In certain circumstances, when the benefits of hiring PSCs outweighs the negative consequences of doing so, however, then using PSCs becomes morally viable—perhaps the Haiti relief effort is one of those cases. The January 12, 2010 earthquake that struck Haiti was a true disaster. Estimates indicate that over two hundred thousand were killed and more than one million Haitians were displaced. Already a country with poor infrastructure and low standards of living, Haiti is and remains unable to either secure her citizens or provide them basis care and services. With regional military and government resources to commit to Haiti relief, the large loss of life, and the potential for increased wide-scale human disaster, one could make the case that the benefits of hiring PMCs to assist with the relief efforts would be a morally permitted option.

As I have mentioned, however, generally speaking, the negative consequences of commodifying forces through the hiring of PSCs counterbalances the benefits. Some may counter that these negative consequences are merely expected and may not occur. This may be the case, but one might point out that so too are the benefits of privatizing security; further, those benefits have historically fallen short of expectations. Clearly then, a serious exploration of the foreseeable risks of hiring PSCs must be undertaken prior to letting a security contract. And, all things being equal, a prohibition against the commodification of force through the hiring of PSCs makes sense. The measurement of costs and benefits is always open to interpretation. But, when the clearly measurable costs savings fail to materialize and when the risks of foreseeable consequences is so great, as is the case with hiring PSCs, states should make it their business to stay out of the PSC business.[55]

Notes

1 This chapter is adapted from David M. Barnes (2013) 'Should Private Security Companies be Employed for Counterinsurgency Operations?', *Journal of Military Ethics*, 12:3, 201–24. Reproduced with permission of Taylor and Francis.
2 For a discussion on how Weber's view has changed in a world of weakening states and more powerful non-state actors see Bailes, Schneckener, and Wulf's "Revisiting the State Monopoly on the Legitimate Use of Force" (2006).
3 Here I assume a positive meaning of a states monopoly of the use of force, i.e., that the people accept the "legitimacy" of the state monopoly. I will turn to a stronger reading of Weber later.
4 *Merriam Webster's Collegiate Dictionary.*
5 See Marx's *Capital* Vol. 1, Chapter 1, Sect. 1 & 2 in Karl Marx and Friedrich Engels' *Capital: A Critique of Political Economy* (1967).
6 This definition is often attributed to Margaret Jane Radin (1996).
7 Cf. d'Errico, (1996), Fox (2008), and de Castro (2003).
8 Prince also advocated using private contractors to fight Boko Haram in Nigeria (Roston 2015). The reader will recognize that Prince in not merely advocating for limited security

and policing functions, but Operational Support services like those provided by EO and Sandline. I will discuss this idea and trend in Chapter 7.

9 See also Carmola (2010).

10 Cockayne further adds, "This system is held together by a complex legal framework controlling violence between those states, in spaces not controlled by any one particular state, and by non-state actors.... In some cases, states have even adopted inter-state liability rules incentivising effective enforcement regimes for the control of non-state actors by state regulators within their own territories" (199).

11 The study of why humans fight in war is rich: see Grossman's *On Killing* (1996), Grossman and Christensen's *On Combat* (2008); Keegan's *The Face of Battle* (1976), and Marshall's *Men Against Fire* (1947, 2000).

12 Singer (2003) writes, "In weak or conflicted states, many multinational corporations see security as just another function that they have to provide themselves, comparable to providing their own electricity or building their own infrastructure. So, in a search for the best protection, the firms, particularly those operating in the midst of civil wars, often hire provider sector military services to protect their investments" (227).

13 Rose also notes that "a second especially important form of inalienability permits transfer by gift but not transfer by sale or purchase" (408).

14 See Sandel (1998 and 2012), Lucas (2005), and Radin (1996).

15 GEN(Ret.) Stan McChrystal (2013) has offered a concept of universal service for young people, and his ideas seem to spring from ideas similar to what Sandel terms the ideal of citizenship. McChrystal is not calling for conscription; rather he is proposing that all could serve their community is some way, like the AmeriCorps.

16 For example, these may include both immediate benefits, including 'reduced expertise acquisition costs, reduced expertise maintenance cost, reduce administration costs, and increased efficiency through specialization,' as well as long-term benefits, such as 'binding multiple principals to a common delegation arrangement, reducing negotiation costs, [and] decision-making efficiencies' (Cockayne 2007, 198). Cockayne also lists blame shifting, 'because agents—not principles—carry the consequences of unpopular decisions.' Although the last may or may not be a benefit overall; I think that it is not one. Blame shifting invokes matters of government transparency, which are worth discussing later.

17 Isenberg (2009) writes that: "The Third Wave had three purposes: (1) to free up military manpower and resources for the global war on terrorism, (2) to obtain non-core products and services in the private sector to enable Army leaders to focus on the Army's core competencies, and (3) support the President's Management Agenda. The Third Wave not only asked what activities could be performed at less cost by private sources, but also asked on what activities the Army should focus its energies" (17).

18 Examples in the current war are Arabic, Farsi, Pashto and Dari linguists. In each of my deployments, our unit used contracted linguists as interpreters. See also Avant (2005). She writes, "PSCs can draw on a deeper pool of personnel with area expertise. In the ACRI program, for example, MPRI was able to provide French-speaking instructors for francophone African states that would not be available from the ranks of the Special of forces" (123).

19 LOGCAP provides all preplanned logistics and 'engineering/construction-oriented contingency contracts and includes everything from fixing trucks to warehousing ammunition to doing laundry, running mess halls, and building whole bases abroad.' LOGCAP has also been used in Iraq and Afghanistan. Currently the Pentagon is operating under LOGCAP 4, and three contractors, KBR, DynCorp, and Fluor Corporation, make up the contract. I have worked with different iterations of LOGCAP overseas.

20 Data from the CENTCOM 2d Quarterly Contractor Census Report; see Schwartz (2009, 5).

21 The ISOA (formally the International Peace Operations Association—IPOA) is a 501(C)(6) nonprofit trade association, whose members are firms 'active in the peace and stability

operations industry.' Great Britain also has a professional organization for their private security companies: the British Association of Private Security Companies (BAPSC). According to their website, BAPSC "aims to raise the standards of operation of its members and this emergent industry and ensure compliance with the rules and principles of international humanitarian law and human rights standards."

22 (Perlo–Freeman and Sköns 2008). Peter James Spielmann (2013) writes, "The private military and security business is growing by 7.4 percent a year and on track to become a $244 billion global industry by 2016 . . . ," and this study was released before ISIS became an adversary and household word.

23 The contractor numbers alone are staggering. As Schwartz (2010) notes, in "the [US'] three largest operations of the past 15 years . . . contractors have comprised approximately 50% of DOD's combined contractor and uniformed personnel workforce" (1).

24 In addition, Cockayne notes that, 'screening costs in selecting agents; negotiation costs, including building and institutional checks into principal-agent arrangements; monitoring costs, whether from police-patrol monitoring (direct monitoring by the principle) or fire alarm monitoring (relying on third-party testimony or whistle-blowing); and the cost of sanctions' (2007, 198) must be taken into consideration in addition to the contract costs and the above secondary costs.

25 Carafano (2008) writes that "In 2005, the federal government issued well over 1.5 million fixed contracts, valued at well over $180 billion. That number represented just under half of all federal dollars spent on contracting" (77).

26 Cf., Singer (2003, 152), Carafano (2008, 77), Avant (2005, 85), and others.

27 Carafano (2008) also notes that cost-plus contracts in the US "more than doubled in the five years after 9/11 to $110 billion. They are used extensively in Iraq. The single largest cost-plus contract is LOGCAP. In 2005 it was worth over $5 billion to KBR, but that hardly meant that KBR made outrageous profits. On average, the company's profits for LOGPAC in Iraq were lower than those for the LOGCAP contracts it fulfilled in Bosnia. Nor was the KBR deal unprecedented. In 2005, the Defense Department handed out three costs-plus contracts for managed health care is totaled almost $6 billion. Interestingly, although the KBR contract garnered many headlines and smart salacious stories, the healthcare contracts attracted scant attention from the national media" (78).

28 See Verkuil (2007, 131). Carafano (2008) also writes: "In the year before 9/11, according to one congressional report, the federal government issued $67.5 billion in sole-sourced contracts, but in 2005 the figure more than doubled to $145 billion. Sole-source and limited-competition contracts are used extensively in Iraq. Many reconstruction projects let in 2003, for example, were limited to a handful of companies bidding for cost-plus contracts worth billions" (78).

29 Isenberg (2009) writes: "In June of 2005, DynCorp, Blackwater USA, and Triple Canopy were awarded contracts under what is now known as the WPPS II [(Worldwide Personal Protective Services)] contract. Personnel qualifications, training, equipment, and management requirements were substantially upgraded under WPPS II because of the ever changing program requirements in the combat environment of Iraq. The current contract was awarded in July 2005. DS utilizes the WPPS II umbrella contract under which it issues task orders to the three qualified companies: Blackwater USA, DynCorp, and Triple Canopy. The bulk of the contractors come from Blackwater. The contract has a ceiling of $1.2 billion per contractor over five years (one base +4 option years). There are currently seven active task orders under WPPS II: Jerusalem, Kabul, Bosnia, Baghdad, REO Basrah, REO Al Hillah, and REO Kirkuk (including USAID Erbil). An eight operational task order for aviation services in Iraq was awarded the Blackwater USA on September 4, 2007, and performance was to begin in late November 2007. Task Order 1 covers the contractors' local program management offices in the Washington DC area" (30).

The $10 billion umbrella WPPS III contract was let by US DOS in 2010 to eight PSC companies, including Aegis Defense, DynCorp, EOD Tech., Global Strategies Group,

SOC, Torres International, and Triple Canopy (FedBizOpps.gov 2010). In 2015, Triple Canopy won a $47.8 million Department of Homeland Security contract to provide security services in Oklahoma and another contract in New Jersey ("Triple Canopy" 2015).

30 Blackwater's name has changed several times over. Blackwater USA changed to Xe Services LLC. Now it is called Academi. See www.academi.com. While this rebranding may seem innocuous, one cannot help but speculate as to the reasons behind the name changes. As one anonymous reviewer thoughtfully commented, PSC names can change, but an army's does not.

31 Singer (2003) also cites another example. He writes, "For example, a common complaint with PSCs landmine clearance operations is that they often clear only major roads (both easier to clear and also more common measures of contract success), and the risky, but still necessary operations, such as clearing rural footpaths or the areas around schools are generally ignored" (157).

32 Cf. Singer (2003, 159); Verkuil (2007, 131); and Avant (2005, 127). As Avant notes, hiring PSCs also reduces incentives for the military to reorganize to fill the service requirements.

33 Nevertheless, there is a risk that the government would be overcharged for the service provided. Custer Battles company representatives accidentally left a spreadsheet behind, which, was later discovered by CPA employees. "The spreadsheet showed that the currency exchange operation had cost the company $3,738,592, but the CPA was billed $9,801,550 a markup of 162 percent" (Isenberg 2009, 87).

34 I thank Ben Hale for alerting me to this issue.

35 In March 2009, according to the CENTCOM 2nd Quarterly Contractor Census Report, TCNs made up 45% of the US DOD hired PMC workforce in Iraq, while TCNs provided 10% in Afghanistan for a combined total exceeding 67,000. (These numbers do not include other non-DOD PMCs.)

36 For instance, Lt. Gen. (Ret.) Jay Garner, who was the Director of the Office for Reconstruction and Humanitarian Assistance for Iraq following the 2003 invasion was protected by former South African military personnel. (Isenberg 2009, 38) Cf. Salopek (2007).

37 Isenberg (2009, 39) discusses that "In May 2005 Honduras's Labor Ministry announced an offer it had been asked to relay from the US firm Triple Canopy, which was willing to pay comparatively high salaries to recruit 2,000 Hondurans to work as security guards in Iraq and Afghanistan."

38 CENTCOM 2nd Quarterly Contractor Census, "Contractor Support of U.S. Operations in USCENTCOM AOR, Iraq, and Afghanistan" (2009). In 3d quarter, FY 2014, CENTCOM reported that there were 3177 PSC personnel in Afghanistan working for the DOD. This number does not include the ~20,400 Afghan Public Protection Force personnel or the 220 Risk Management Consultants (2014).

39 Avant (205) discusses a different example of "Reports that MPRI was training the Kosovo Liberation Army (KLA) in the midst of the war in Kosovo—vociferously denied by MPRI—were due to just this dynamic. Though MPRI's claims were true, persons who had worked for MPRI at one time also did freelance consulting with the Albanian government and may have provided services to portions of the Kosovo resistance" (222).

40 Cf. Isenberg (2009, 21); Singer (2003, 211–14); Avant (2005, 4); and Silverstein and Burton-Rose (2000, 143).

41 The Russians are an example. See "Bill to Allow Private Military Contractors Submitted to Russian Parliament" (2014). The article quotes: "'*We must also create something that could use our military veterans who are ready to execute military and security tasks, including abroad*,' Nosovko added."

42 For discussion of warlordism, see Kinsey (2006, 112–15). He writes, "Taken to its logical conclusion, warlord politics sees collective and private authorities resembling one another, but with the emphasis on different values. For example, each type of authority

provides security. In the case of collective authority, security is a right of membership to the state, while in the private sphere security is a consequence of other factors, most notably economic interests" (115).

43 Perhaps ironically, shipping companies are now turning to PSCs to protect their ships from piracy, often encouraged by both insurance companies and the state navies. See the Center for International Maritime Security's (CIMSEC) *Private Military Contractors: A CIM-SEC Compendium* (Hipple 2015). This compendium looks at the history and policies surrounding the use of private maritime security companies. Also see Cheny-Peters (2014), Isenberg (2012), and Pitney and Levin (2013).

44 For a discussion of the violence against civilians, see Stepanova's "Trends in Armed Conflicts: One-Sided Violence against Civilians" (2009).

45 Singer (2003) raises another concern. He writes, "The shift in capabilities can affect assessments of power, and the balancing that results, in numerous manifestations, including the potential increase in miscalculation. Possibly engendered by a firm overselling his services, a client could develop a misplaced belief in the dominance of their reinvigorated offense and initiate war when they are actually not at an advantage. This was a concern with contracts of military consultant firms in the unstable Balkans" (175).

46 "'We are loathed out here. We are the single most hated entity in Iraq,' said Ethan Madison, a security contractor who has worked in Baghdad for five years." Judd (2009).

47 Verkuil (2007) discusses the Bush administration's choice of using the TSA in great detail (57–71).

48 Some might argue that the UN and other regional organizations (e.g., EU or NATO) may one day assume responsibility for the use of force. Some have even argued (as I have explored elsewhere) that only a global, representative organization might truly be considered a jus ad bellum legitimate authority. Nevertheless, the monopoly of force would continue to exist; it would merely be held by the states' replacement.

49 I thank David Boonin for raising these concerns.

50 Verkuil (2007) writes that "In both the IRS and Coast Guard examples, the outsourcing of management functions that are best performed in-house undermines government performance in two ways: by utilizing second-best performers and by weakening or atrophying government's power to perform these functions in the future. Government managers cease to exist when they are not put to good use" (4).

51 Boonin also raises this issue a different way. He asks, "What are government or military functions? When certain other non-security functions become necessary for carrying out security functions, can these other functions be outsourced, and if so, then how are they different from the PSC services?" The US government categorizes certain functions as inherently governmental. Isenberg (2009) notes that "The Office of Management and Budget [(OMB)] lists the following functions is inherently governmental: interpreting and executing laws; ordering military or diplomatic action on behalf of United States; conducting civil or criminal judicial proceedings; performing action that significantly affect the life, liberty, or property of private persons; and collecting, controlling, or disbursing appropriated and other federal funds" (117, footnote 28). Much as I think the consequentialist case can be made for privatizing logistics to some PMCs, contracted army doctors, for example, serving in a necessary, non-security function should be allowed.

52 I thank Eric Chwang for introducing this analogy.

53 Cf, the Institute for Security Studies: Sierra Leone, and Jäger and Kümmel (2007, 107).

54 For a different perspective highlighting other ethical consequences of hiring PSCs, see Marble-Barranca (2009).

55 I would like to thank the audiences at the 2010 International Society for Military Ethics (ISME) and the 2010 Zentrum für ethische Bildung in den Streitkräften (zebis) conference, Alastair Norcross, David Boonin, Ben Hale, Eric Chwang, George Lucas, Richard Schoonhoven, and Martin Cook, as well as the anonymous referees for *Journal of Military Ethics*, for their comments on earlier versions of this chapter.

5 The belligerent equality of armed contractors?

"With your permission, sir, I have acquired a target."
(Cpl. Lonnie Young in Prince 2014, 137)[1]

On April 4, 2004, hundreds of Shia militants attempted to storm the CPA compound in Najaf, Iraq. The fighting was intense and lasted over four hours. Dozens of the attackers, even hundreds by some accounts, were killed or wounded, and the defenders of the compound suffered three wounded. What is most interesting about the battle account was the make-up of the defending force. It was a hodge-podge of eight Blackwater contractors, a few El Salvadoran soldiers, and a couple of US Marines (129–51). One of the Marines, Cpl. Young, who was there to work on communications equipment, played an integral part in the battle. Armed with an M249 squad automatic weapon (SAW), he provided covering fire, evacuated wounded, and carried ammunition for resupply until he was evacuated in a Blackwater "Ass Monkey" Littlebird helicopter (Scahill 2007, 117–32). When he requested permission to fire, to engage the attackers, he received permission not from his superior officer, but from a Blackwater contractor. As Erik Prince notes (2014), this was the "first time in memory that private contractors commanded active US military personnel—in the midst of a firefight ..." (138).

The in bello tenet of discrimination, and its legal equivalent distinction, is on its surface a rather simple concept concerning conduct in war: According to discrimination, one is only allowed to deliberately target certain individuals (or structures). Correspondingly, there are others whom one may never intentionally target. In international law and the LOAC, we think of these groups as combatants and non-combatants, respectively. And, while it is conceptually simple to distinguish between the two groups, theories concerning how to categorize them and to what extent vary, and in practice, in the heat of battle, identifying permitted targets from prohibited ones is never easy.

A related principle in Just War Theory to combatant-non-combatant discrimination is the MEC:

> **The Moral Equality of Combatants (MEC):** Combatants on all sides in a war have the same moral status. They have the same rights, immunities, and

liabilities irrespective of whether their war is just. Those who fight in a war that is unjust ("unjust combatants") do not act wrongly or illegally when they attack those who fight for a just cause ("just combatants"). They do wrong only if they violate the principles governing the conduct of war.[2]

Combatants, then, are morally permitted to kill, but they also may be targeted and killed by their opponents. But, what about private contractors? In particular, how do we morally and legally categorize armed contractors like those who fought in Najaf, and does the MEC apply to them as well?

In his article, "On the Moral Equality of Combatants," Jeff McMahan (2006c) argues against the Just War Tradition (JWT) convention separating jus ad bellum and jus in bello or the decision to go to war and conduct during war.[3] In particular, he rejects the convention of MEC. McMahan (2006a) thinks that this equal moral status is false. He holds that the rules that govern the permissible behavior of soldiers fighting on the just side of a war are substantially more permissive than the rules that govern the permissible behavior of soldiers fighting on the unjust side of a war. Although it is permissible for soldiers on the just side to intentionally kill soldiers on the unjust side, it is not permissible for soldiers on the unjust side to intentionally kill soldiers on the just side (379). Further, his conception of the moral *inequality* of combatants seems prima facie correct, especially when looking at some historical examples. For instance, would one want to acknowledge that the Nazi soldiers were the moral equivalent to the allied soldiers during World War II? The issue of belligerent equality becomes even more complicated when considering the moral status of children soldiers, insurgents, or even terrorists.[4]

Nevertheless, while problematic, and I think it raises some tough questions, MEC is true. Furthermore, I think that MEC helps draw distinction between the moral status of armed contractors and of professional soldiers. It is the very conception of MEC that highlights this distinction in moral standing and what drives much of our moral intuitions and unease over the employment of armed contractors. Part of the issue at hand is that McMahan and others have focused attention on whether combatants can be just in an unjust war.[5] This is an important question and may shed light on a soldier's obligation to refuse to go to war; that is, conscientious objection or disobedience. Furthermore, limiting a justified status to combatants in only a just war may help limit self-proclaimed just soldiers' participation in unjust wars, even limiting prosecution of an unjust war itself. Certainly, these are noble aims; however, I think that McMahan simplifies this idealistic solution, and more directly he overlooks the deeper problem.

McMahan categorizes combatants into two groups—just and unjust combatants. These two groups are further separated from the innocuously-named larger collection of non-combatants, which of course includes the civilian population (and as Congress has maintained—contractors) (Elsea, Schwartz, and Nakamura 2008, 2; 14). The question of determining combatant status plays an important role in the jus in bello tenet of discrimination, and it will show that while armed contractors are acting as combatants (certainly they seem to be by their actions *not* non-combatants), they are not combatants with the same moral and legal status of professional soldiers. This problem is magnified because Congress has repeatedly categorized PMC contractors (including PSCs) as civilians. Moreover, this moral distinction extends beyond mere legality or McMahan's just-unjust distinction. There exists a MEC between soldiers that does not extend to armed contractors.

First, I will outline McMahan's characterizations of MEC proponents' arguments for MEC, agreeing with him that several of these arguments are too weak to support MEC. After mentioning other arguments for MEC that McMahan either does not discuss or does not adequately refute,[6] I will then present what I believe to be the strongest case for MEC. Once I have argued for MEC, I will then turn to the question of the belligerent equality of armed contractors. By focusing on defining what it means morally to be a combatant, I will argue that armed contractors are not the legal equals of combatants nor are they the moral equivalent of professional soldiers. The Blackwater contractors in Najaf were not the legal or moral equals of the Marines, nor were they civilians. I offer that legally and morally they represent something else.

McMahan's argument against MEC

MEC can be viewed in terms of two co-related theses that Rodin aptly calls the Symmetry Thesis and the Independence Thesis (2007, 591). The Symmetry Thesis refers to the rights, liabilities, and obligations of combatants in terms of jus in bello; for soldiers, these remain the same. The Independence Thesis recognizes the separation between the jus ad bellum, or the justice of the war itself and the jus in bello—concerning the actual fighting of the war. McMahan denies the Independence Thesis, which in turn blocks the Symmetry Thesis. While McMahan is not the first theorist to question MEC (See Holmes 1989, especially pages 176 and 186), he is the first to point out that Walzer's (1977) formulation of MEC actually shows that our intuitions of both the Symmetry Thesis and the Independent Thesis may be incorrect.

Discussing the MEC, Walzer (1977) wrote,

> When soldiers fight freely, choosing one another as enemies and designing their own battles, their war is not a crime; when they fight without freedom, their war is not their crime. In both cases, military conduct is governed by rules; but in the first the rules rest on mutuality and consent, in the second on a shared servitude.
>
> (37)

McMahan (2006c) interprets Walzer as arguing that combatants are equal to either their entering combat as mutually consenting contestants, such as boxers,[7] or as mutually coerced participants, such as gladiators.[8] McMahan raises objections to both of these views of combatants to show that a defender of MEC would need a stronger argument.

Proponents of MEC often turn to contractual based arguments[9] or to a soldier's commitment to the military[10] as reasons for her moral equality. McMahan thinks that these also fail to support both the Symmetry and Independence Theses.

Furthermore, while McMahan accepts that often soldiers cannot be completely knowledgeable of whether a particular war is just, to him it is reasonable that most combatants have access to at least some information as to whether the war they are fighting is just. Defenders of MEC often raise this epistemological problem as grounds for accepting MEC.[11]

McMahan is right to note that each of these proposals for MEC are insufficient, and I must address what I think is his most persuasive argument. In his own words, "it's hard to see how any means to the achievement of an unjust end could be anything other than wrongful" (2006c, 379). In other words, any jus in bello act in an unjust war, regardless how closely it follows the principles and laws associated with proper conduct of war, is instrumentally benefiting the unjust war.

McMahan's most persuasive argument

Another way to look at McMahan's response is that a combatant's actions in an unjust war, while they may have good consequences in a limited scope (i.e., within the jus in bello), they will have the broader consequences of enabling or contributing to the unjust war. McMahan's argument is a consequentialist one. Furthermore, he argues that, "Conscientious refusal to fight in an unjust war can reduce the probability that the war's unjust ends will be achieved, hasten the end of the war, thereby reducing the number of casualties, and contribute to the deterrence of further unjust wars" (2006c, 386). Unjust combatants refusing to participate in an unjust war might then result in better overall consequences. Certainly, this appears to be the case. However, one might note that a just combatant who uses unjust means might do so to end the conflict sooner. But, one could use the same line of reasoning to argue that any harm done in an unjust war has negative consequences, even the harm committed by soldiers on the just side, especially if the unjust side wins. One might take this view to show that there should be no war (i.e., pacifism). However, while I do agree that a world without war would be better, wars, just or otherwise, seem highly probable.[12]

McMahan would counter that because there is a greater than fifty percent chance that a combatant will end up fighting for an unjust side,[13] the moral risk to the combatant seems to point to her instrumentally assisting the unjust cause. Nevertheless, this analysis does not seem to take into account that the combatant's moral risk is a combination of probability and the weight of the consequences. Even assuming that for any combatant, there is a greater probability of taking part in an unjust war, the consequences of fighting—should the war be a just war—might outweigh the consequences of sitting out a possible unjust war.[14]

Finally, as I will return to later, there seem to be practical reasons for a legal equality. McMahan (2007) agrees, writing that while he believes "most combatants who fight in unjust wars thereby act impermissibly, that most nevertheless have excuses that mitigate their culpability and are sometimes entirely exculpating ...." While seemingly merely pragmatic, and legalistic, it is these so-called "pragmatic" reasons that very real and moral consequences of moral equality. There is not only a pragmatic, legal equality between combatants; MEC is true because of the better overall consequences, including the consequences of following the LOAC and in bello principles, restricting the destructiveness and brutal nature of war.

Civilian (non-combatant) immunity

Furthermore, if, as McMahan argues, MEC is false, then the question of in bello discrimination or who can be legitimately targeted comes into question. In his "Civilian Immunity and Civilian Liability," McMahan (2009) acknowledges this possibility. If civilians can be liable for war—even in a limited way, then in principle civilians may become liable to be harmed in war. And, in certain circumstances, they can become legitimate targets (210; 212). And, while we may be less concerned with PSCs as civilians being targeted than innocent civilians, it is because of our intuitions that non-combatant immunity exists.

While McMahan acknowledges that a legal prohibition against targeting civilians is preferable, he asserts that it is not based upon a non-combatant status or any moral status (which makes one wonder about the grounds for a legal immunity). Nor, McMahan argues, is it based upon a weighing of the consequences, which is false because the reasons for a legal prohibition certainly seem to reflect the consequences resulting from targeting civilians. He notes that it is a combination of a variety of reasons, including: (1) a low degree of individual responsibility; (2) attacks against civilians uses them as mere means; (3) attacks against civilians have been certain effectiveness; and (4) the guilty and innocent are intermingled, and so on (231). To me, however, these seem sufficient to not only maintain a legal discrimination (distinction), but they capture the moral grounds of civilian immunity and in bello discrimination that McMahan seeks. In other words, McMahan's reasons reflect the consequentialist argument for why MEC is true and why non-combatant immunity is morally justified and ought to be legally codified. To argue legal equality and legal distinction without moral grounds seems arbitrary at best. Additionally, the better state of affairs where non-combatant immunity is true seems to strengthen the Symmetry Thesis of MEC.

Partial solutions to the MEC problem

Before I outline this solution, it is important to mention that there are possible alternative counter-arguments to McMahan's objection to MEC. Each of these proposals approaches the MEC problem in a different way. The first, offered by Benbaji (2008c), uses a reformulation of J. T. Thomson's self-defense argument to counter McMahan.[15] Bomann-Larsen (2007), alternatively, views combat according to the MEC as fighting between justified and exculpated aggressors.[16] The third is the collectivist solution, which attempts to refine the argument for MEC from a division of moral labor approach, informed by a combatant's institutional commitments.[17] I think that while these proposals are not fully adequate, they do provide partial solutions that both identify weaknesses in McMahan's argument and provide strong reasons for MEC.

An argument for MEC

> The most obvious form of rejoinder to the asymmetry arguments is to claim that they must be rejected because they would lead to disastrous consequences if they were ever to be implemented in a working regulatory regime of war. We might ... [refer] to this from of objection as the "pragmatic" response to the arguments for asymmetry, while recognizing that these objections do not necessarily reflect a theoretical commitment to utilitarianism or consequentialism more generally.
>
> (Rodin and Shue 2008b, 7)

Rodin and Shue, by their choice of words, appear reluctant to accept a "pragmatic" or consequentialist solution. This is, however, exactly what I intend to do. I argue that MEC is true, with its symmetry and independence components, as a result of its overall better state of affairs than one where MEC does not exist.

Holmes (1989) writes that, "Whatever the justice of one's resort to war, there are in this view limits to what one may do in conducting it" (176–7), and I agree. In addition, acknowledging combatants as moral equals, not only has more pragmatic benefits in the legal sense (as McMahan discusses), but also provides the moral basis for the key restrictions of jus in bello, namely to limit the harm and destruction inherent in war, foremost by noting whom may be killed and who or what can do the killing.

Even Walzer (1977), while he does not accept so-called unconstrained "utilitarian calculations" in matters of in bello, notes that "every war in a series extending indefinitely into the future were to be fought with no other limits ..., the consequence for mankind would be worse" (131).[18] Bellum justum, informed by MEC, provides a set of prohibitions and limits on total, unrestricted war. Hurka (2008), although a proponent of a deontologist view of war, thought that "a morally crucial fact about war is that it causes death and destruction" (127). McMahan (2006c) should concur with me on this point; he writes that "War has such serious consequences ... that it must be subject to institutional constraints designed to ensure that it's not undertaken without moral justification" (392).

I will show that a world where MEC is true is consequentially better. In particular, I will show that: (1) both just and unjust combatants can ignore or violate the LOAC and perpetrate war crimes, and we want to condemn wrongful conduct in war separately from whether the decision to go to war was just. MEC allows us to do just this; (2) allowing that only combatants fighting a just war (with a just cause) can rightfully kill—in this case—unjust combatants, raises the ugly concern of an inability to limit the destructive nature of war (e.g., dehumanizing the enemy to wars turning into crusades to an easing on the restrictions of military necessity), which is the point of bellum justum; (3) as I mentioned, what defines combatants as combatants (whether just or not) remains murky at best and it seems that just cause alone does not entail just combatants; there are other conditions to be met; and (4), removing MEC threatens non-combatant immunity, both in theory and in practice. Both (1) and (2) strengthen the case for maintaining MEC; while addressing concern (3) leads to defining who are combatants. Should abandoning MEC entail (4), then the undermining of non-combatant immunity turns the traditional conception of a morality during the conduct of war on its head.

Pragmatic independence

Recall that the Independence Thesis states that the in bello realm is separate from the ad bellum. Additionally, there have been many reasons put forward for maintaining this independence. I will present what I think are the strongest cases for independence—each one contributing to the overall better consequences of the Independence Thesis and MEC.

Ad bellum justice may be mixed[19]

In bellum justum, ad bellum is traditionally viewed as encompassing six tenets that must be satisfied for a state to justifiably go to war: legitimate authority, (macro) proportionality, last resort, public declaration, reasonable chance of success, and just cause.[20] Just cause is arguably the strongest, most prominent of the tenets, but it is also one of the most controversial. On the one hand, one could argue that as long as all tenants are met, war is just. However, there are several potential problems here, each of which call into question the reliability of a combatant having epistemological certainty of her side's cause. McMahan (2008) even acknowledges that there is uncertainty about just cause, writing: "even the supposed experts disagree about the justice of particular wars" (27). The problem of determining whether a cause is just is threefold. First, the justice of a particular war can be mixed. Second, there can be more than one cause for war—a case of plural causes. Third, the cause, or rather what constitutes the just cause, may change.

That a war can have mixed causes should not strike many as controversial. Overthrowing a tyrannical regime, instituting democratic governance, and securing access to natural resources may all be intertwined into an overall cause or may be used as a menu of just cause convenience. This mixing of causes seems especially true in cases of conflation between the tenets of just cause and proportionality. A related issue is the problem of plural causes.[21] Furthermore, as McMahan observes, some of the missions conducted by the soldiers on a side might support either the just or unjust plural causes (31).

Assume, however, that there is only one cause—a just cause—at the start of the war; it is the reason why the state went to war. If the cause were to change during the conflict, would it call the whole war's justice into question? What about the benefit of hindsight? Consider the discovery of the concentration camps towards the end of the Second World War. Each of the problems of mixed causes, plural causes, and changing causes point to the contentious problem of determining whether a war is just or not. If we as academics and people with time to reason and argue over whether any wars are just cannot agree, then why should we expect the combatant could do any better? As Rodin (2007) observes, "War is so difficult, so dangerous, and so costly, that it is exceptionally difficult for ordinary humans to undertake it without believing that they are in pursuit of a cause that is noble and just" (602).

Ignorance of jus ad bellum

Even if one were to suspend the problem of ad bellum justice, the question remains over whether there should be an expectation that combatants have access to these facts and also whether the lack of this access should be morally relevant. It seems reasonable to assume that most combatants believe that their cause is just; it has even been suggested that it would be somehow unfair to combatants to hold them accountable for either the justice of the war or their belief in its justness. The epistemological defense cannot rely upon a conception of false belief eliminating one's moral responsibility. No, it is rather a combination of: (1) a personal, genuine belief in a cause's justice; (2) a lack of epistemic certainty—an ignorance—of the justice of a war; (3) the institutional-based ignorance of combatants; and, (4) the institutional commitments of the combatants that point to the theoretical and practical separation of jus in bello from jus ad bellum.

Sidgwick (1919) asserts that "we must treat both combatants on a subject that each believes himself in the right" (253). In addition, Bomann-Larsen (2007) contends, "in a democracy there are pro tanto reasons, not just prima facie reasons, to respect the democratic division of labor and respect the outcome of a democratic decision-making procedure—even if they result in the wrong decisions" (177). There are others, politician leaders for example, who are responsible for making ad bellum decisions. Furthermore, it is the very nature and structure of the combatant's institution that contribute to her ad bellum ignorance. This seems especially true in not only totalitarian states, where information is closely controlled, but also in democratic states. Yet, the ignorance of ad bellum is not only informationally based, or even institutionally bred. A combatant's worldview, especially for one who has served within the institution for long periods of time, creates an environment where ad bellum issues move outside of her purview.[22]

When one overlays the problem of ad bellum justice to a combatant's practical ignorance of a war's justice, it seems clear that the independence between the two levels of war remains. War, Rousseau (1893) writes, "is not a relation between man and man, but a relation between state and state, in which individuals are enemies only by accident... as soldiers" (14). The combat itself is fought between individuals or groups of individuals, but war is as much a political endeavor, fought for political aims. The Independence Thesis of MEC reflects this very real separation.

Preventing private wars

Another reason for evaluating the justice of war is to limit non-state violence and prevent private wars. The threat of increased non-state violence and the recent wars in Iraq, Afghanistan, and Syria, as well as a resurgence of piracy, call into question how the international community should treat non-state combatants; indeed, are they combatants at all? Nevertheless, combatants, according to the LOAC, are members of their military institutions.[23] Not all military are of course legitimate, nor are all their methods just. However, there is an appeal that if there will be war, it is limited by constraint on the number of political institutions participating. By reducing private violence, theoretically, controls can be established over war. As Zupan (2008) points out, "there does seem to be a real problem if [soldiers] were given a lot of latitude about the wars [they] fight" (225)[24] and starting one, when a soldier believes it to be just, highlights this problem and its absurdity. Recall that this is a concern for those who believe that PMCs offer a way for governments to fight proxy wars, even private wars.

If one maintains that a soldier is responsible for the ad bellum decision to go to war, is she correspondingly obligated to go to war when the cause is just, or is it merely permitted? Consider a different case: might this line of reasoning also imply that she should initiate just wars. For example, what if a soldier or a group of soldiers believes that a particular cause is just, even when their state refuses to act? Should they take it upon themselves to do so?[25] While this example may seem far-fetched, it is useful to show that if combatants have a responsibility for ad bellum decisions, then this responsibility may only go in one direction. The legitimate authority tenet of bellum justum restricts, Rodin and Shue (2008b) note, "the right of initiating war to properly constituted sovereign bodies. This limitation of war-making authority has undeniably been a real moral achievement" (13). Maintaining the decision to go to war at the political level, therefore, seems strong evidence for the Independence Thesis.

Obedience and deterrence

Part of the argument against the Independence Thesis points to a soldier's obedience to the state and its military as counterproductive to the prevention of unjust wars. What this argument fails to acknowledge, however, is that it is in part due to this same obedience of the combatant to her institution that supports deterrence against unjust attacks from other states. In filling the role of the soldier, while it is true that a person gives up some of her autonomy, she also inherits certain obligations to her state.[26]

However, on entering the military she must consider her military's and government's past actions. Maple (1998) writes, "Both volunteering and allowing oneself to be conscripted into a standing army should be considered permissible or blameless unless the state has a clear record of using force unjustly." This is partly how one exercises moral autonomy in the military. However, when the risks of becoming an unjust combatant "are significantly outweighed by political obligations to support the state, volunteering for a standing army may even be morally required" (184). Moreover, it is hard to see how this same moral obligation extends to the private contractor, when she is only bound by the contract itself.

Furthermore, if we assume that the military is a legitimate institution,[27] then the soldier is a legitimate member and owes her obedience to the lawful commands, commands to go to war and lawful orders when waging war. If, however, the military (as a legitimate institution) is used for unjust political causes, then it does not necessarily call the military institution's, or the soldier's, legitimacy into question as well. Unjust combatants as well as just combatants are both lawful combatants, not outlaws. It is their unlawful and immoral conduct that would make them so. In fact, it is because of the Independence Thesis that one can judge the actions of combatants separately from the political decisions to go to war.

The existence, then, of an obedient and effective military, Benbaji (2008c) writes, also "minimizes the extent to which soldiers and civilians are vulnerable to unjust threats" (490).

Double jeopardy

Walzer (1977) succinctly captures the idea behind the Independence Thesis when he writes, "War is always judged twice, first with reference to the reasons states have for fighting, secondly with reference to the means they adopt" (21). With the Independence Thesis, one can evaluate whether a just war is fought with just means or not, and one can separately look at the means used when a war is unjust. According to the separation of ad bellum and in bello, therefore, one judges a combatant's conduct in war apart from the political decision to go to war, and we do so because ad bellum is independent from in bello. For example, one might condemn certain actions of an Allied soldier fighting against Nazi Germany. If the Independence Thesis is an illusion, however, then the combatant is a form of double jeopardy. If the in bello conduct is not separate from the ad bellum decision to go to war, then the combatant can be guilty of both transgressions. It may seem reasonable to do so because the combatant maintains her role as a citizen as well as her role as a soldier.

It seems odd, however, to say that she is guilty of her state's pursuing an unjust war. Let us return to the benefit of preventing private wars. Part of the contractual agreement a soldier enters is a relinquishing of her ability to privately execute justice, even violently. Justice and violence are now state functions. Yet, by denying Independence, one still holds her accountable for the functions she has given up. Zupan (2008) observes,

> It is as if we demand of the individual that she refrain from certain activities (private wars) and seed that authority and responsibility to the state, but at the same time we reserve the right to condemn her for fulfilling the terms of the contract: she is to be in the state of nature and out of the state of nature at the same time.
>
> (216)

Therefore, if we accept that states provide efficient protection from external threats and limit individual violence as well, then it seems that the individual combatant is not legally or morally accountable for the political decision to go to war.

The symmetry-independence overlap

No basis for liability

Rodin (2007), a proponent of both ad bellum dependence as well as a form of asymmetry (restrictive asymmetry), raises an interesting observation relating to the liability of combatants. He notes that a just combatant, one who fights for a just cause, assuming that they were following jus in bello, "seemed to lack liability for force being used against them" (539). Just combatants are permitted to use force, Rodin argues, while combatants fighting an unjust war are not. Thus, the unjust combatants not only "do not enjoy a symmetrical right to kill," but should be held liable post bellum.

Rodin's observation seems to point to the justice of the war as the liability-assigning feature for the combatants. However, this view raises some potential problems. First, as I have discussed, determining the justice of war is unreliable, perhaps indeterminate. Therefore, assigning liability also proves troubling.[28] Second, if only unjust combatants (assuming all sides are in bello-following) are liable, then this poses an unfair burden upon them. Both sides of the war already operate with not only the belief that their own cause is just but also the belief that the other side must be unjust. Add the proscription of liability upon the opposing side, thereby denying them in bello reciprocity, there seem no grounds, moral or otherwise, for states to act in accordance with international humanitarian law.[29] The Independence Thesis and Symmetry Thesis together then maintain issues of the liability of fighting an unjust war at the political level, and allows the clearer assessment of liability of wrongful conduct in war.

War crime prosecution

Separating the question of liability at the ad bellum level from the liability of combatants' conduct also allows for assignment of blame when individuals ignore or violate in bello principles, but more importantly assists in the prosecution of war crimes. Recall that bellum justum offers a counter to the realist position of *Inter arma silent leges* (in war the laws are silent). Both defenders and opponents of MEC would agree that of all acts of war, some seem morally unacceptable on all sides. Regardless of which side is liable for an unjust war, even in cases where both sides are ad bellum liable, the Independence Thesis ensures that one can evaluate conduct on the battlefield. Soldiers on both unjust and just sides are not only entitled legally and morally to disobey orders that "manifestly violate the rules of conduct in war; they, in fact, ought to do so" (Benbaji 2008c, 490). Even just combatants cannot cite civilian orders from their government to prosecute a war with illegal means.

In order then to assign blame or prosecute individual crimes of war, both which may help limit the destructiveness of war, the combatants must be treated as moral equals, and according to MEC, they are. Combatants fighting an unjust war can and should be praised for fighting clean in accordance with jus in bello. Furthermore, as Bomann-Larsen (2007) writes, "A war criminal (for instance, one who violates the principle of non-combatant immunity) seems to be a war criminal regardless of which side he is on. How can we single out war criminals in bello if we deny the validity of the Independence Thesis?" (125–6).[30]

I believe that we cannot. It is both the Independence Thesis and Symmetry Thesis that enable war crimes prosecution. Is not only the mere threat of prosecution that restricts conduct in war, but it also forms the basis for educating combatants on their responsibilities in war, and it informs the international standards.

Consequential symmetry

The argument so far has focused on the consequential reasons for the Independence Thesis as well as arguments for both the Independence Thesis and Symmetry Thesis. In this next section, I will focus on issues particularly related to the Symmetry Thesis, including the problem of victor's justice, compliance, and moral zeal, as well as the effectiveness of current laws involving belligerent equality. Together these arguments will show why MEC exists.

Victor's justice

A very real problem in the prosecution of war crimes is that there has historically been a lack of an impartial arbiter of justice. Although much of the current evolution of LOAC is attributed to the Nuremberg trials after WW II, these trials have been labeled as cases of victor's justice. The winner gets to decide the legal fate of the defeated. Even defendants in the international courts for the Former Yugoslavia have argued that the trials could not provide blind justice. How then could combatant liability be assessed? Furthermore, as McMahan (2008) notes, the "practice of post-bellum punishment of unjust combatants" (30) could actually result in protracted war, as neither side desires defeat, let alone legal prosecution on a part of the victors. A combatant fighting for her state, whether her side is just or not and regardless of whether she believes it to be so, has no incentive to surrender or accept defeat if there remains a constant fear of legal prosecution if her side loses.

Reduced compliance or incentive

Consider the same combatant who fears surrender and defeat; what must one say about her incentive to follow jus in bello at all? Rodin (2007) ponders, what about the "possibility that restrictive asymmetry might reduce the likelihood that unjust combatants would comply with important current in bello prohibitions" (604–5)?[31] While Rodin does not believe that there is a problem with reduced compliance or incentive to follow in bello rules, I think that he is mistaken. The threat of being treated as a war criminal should her side lose, combined with the belief that since her side is just and her opponent's is not, may lead her to accept the realist position that any act that she commits to quickly end the war would be justified, a point McMahan (2008) also seems to acknowledge.[32]

First, as Rodin (2007) writes, "If there is no moral distinction between harming just combatants and harming non-combatants, then there is little incentive for unjust combatants to abstain from the latter given that they are already committed to attacking the former" (605–6). Second, the lack of individual incentive calls into question the broader, institutional incentive to practice in bello as well as to enter into international agreements such as the LOAC. Finally, Bomann-Larsen (2007) points out that to deny belligerent equality not only entails that combatants cannot fight for an unjust cause, but they also cannot defend themselves "against attacks on their persons" by just combatants (24, footnote 26). Since one may also not know whether her side is just, then why fight fair? She must because there is not only a legal equality between combatants, they are morally equal.

Ex ante agreement

For jus in bello to work, for it to be effective in limiting the brutal nature of war, there must be agreement on the part of all parties, and this agreement must occur ex ante.[33] Without the Symmetry Thesis, even a legal ex ante agreement becomes untenable. This agreement "obliges just combatants to treat their enemies as if they are not even minimally culpable for the injustice of the war they fight," Benbaji (2008b) observes. "Additionally, the agreement obliges soldiers to treat civilians as if they are innocent bystanders, whatever their political commitments and contribution to the military effort may be" (12). Combatants enter this "contract" because they believe the rules of the contract fair, mutually beneficial, and commonly followed. Practically speaking, they accept jus in bello because they will not be prosecuted for the decision to go to war, and they also understand that should they become POWs, they will gain certain other rights. Combatants believe this on the grounds of an accepted legal belligerent equality and also because they believe they are morally equal.

Following the current laws on MEC is effective

Even detractors of MEC acknowledge that the rules of jus in bello are effective. Fletcher writes,

> The reason for adopting a rigorous distinction between jus ad bellum and jus in bello is the need for a bright-line cleavage that is workable in the field of battle. Combatants do not have to think about who started the war. They know that, whoever started it, certain means of warfare are clearly illegal.
>
> (quoted in McMahan 2008, 34)

For example, McMahan discusses the successful prohibition against the use of poison gas (41). Others, such as Roberts (2008), point out the effectiveness of the rules of war, in particular the effectiveness of their moral symmetry.

Of course, an objector might point to the actual violations of jus in bello to show that the rules are ineffective. In the Vietnam War, the My Lai incident seemed to typify the apparent failure of the rules of war. More recently, the mistreatment of prisoners at Abu Ghraib and the 2005 Haditha incident, where 24 Iraqis were killed, seem to point to a continued lack of progress. While it is difficult to assess how effective they are, accepting and promulgating the rules of war have had a positive impact, and there are many cases where it would appear that the rules of war are working.[34]

Even if combatants followed the rules of war from purely self-interested reasons, they are still, nonetheless, effective.[35] Restraining wrongdoing and harming in war is consequentially better than unconstrained war; the rules of war grounded in moral belligerent equality are how this restraint is manifested. However, to be effective these rules must apply to both sides and be accepted by both sides—just or otherwise. The in bello rules, stemming from the Symmetry Thesis have been effective and can be helpful in restricting the destructiveness of war. If all combatants are "more inclined to avoid targeting civilians, to respect the rights of prisoners of war, to halt the killing when the other side surrenders, etc." (Benbaji 2008b, 5), then though war may be hell, its scope and devastation may be lessened.

"Moral agents"

One of the arguments that McMahan (2008) uses against the collectivist defense of MEC is that even when the combatant is serving the state, she does not lose her entire moral autonomy. Thus, McMahan argues, she is still accountable for her participation in unjust war. I think that McMahan is partially correct here. I too argue that the combatant does not lose her autonomy; however, it is through the Symmetry Thesis that a combatant retains her equal moral autonomy. Furthermore, there are times where it is morally permitted for unjust combatants to attack just ones, such as when the only way unjust combatants could prevent just combatants from using unjust means. Yet, this supports (1) the idea that combatants retain their autonomy, (2) this autonomy is shared by combatants on both just and unjust sides, and, as I argue, (3) this shared autonomy demonstrates their moral equality.

Combatants may sometimes see themselves as victims of higher politics and fate, but they also, as Rodin and Shue (2008) write, "view themselves in part as moral agents responding to a call to arms which has real moral force" (127). Take away the rights and obligations of unjust combatants, and they lose some of their autonomy. But this does not seem entirely true.

Even when coerced to fight, though they may lose some autonomy, combatants remain moral agents.[36] Recall that anti-mercenary laws codified who has the right to kill and under what circumstances. Though these laws intended that soldiers be licensed, fulfilling a role for the state by acting as agents for the state, they were and are all individually liable for their conduct as autonomous moral agents. In this sense, all combatants are morally equal.

Excessive moral zeal

Walzer thought that one of the benefits of jus in bello was to prevent warfare from becoming one of desert and punishment at the level of the soldier. While punishment may have been a historical just cause, unrestricted warfare at the in bello level, however, could result in a moral crusade by the dominant side. If unjust combatants are not granted the same moral agency, then not only is there less incentive to follow the in bello rules but the perceived unjust enemy also becomes something less than an equal, something less than "poor sods like me" (Walzer 1977, 36). There are many examples of dehumanizing an enemy in war. Even today, labels such as "raghead," "Haji," or "skinny" are widely used (though officially discouraged) because it perhaps gives one less pause when she has to pull the trigger.[37]

Nevertheless, some may counter, as William Paley did in the eighteenth century, that "if the cause and end of war be justifiable; all the means that appear necessary to the end are justifiable also" (quoted in Holmes 1989, 176).[38] Nevertheless, other than the obvious harms of unrestricted warfare, just necessity breeds resentment and desire for revenge—two things that the in bello rules are intended to curtail. As Coates (2008) writes, the "more war is justified, the less restrained it seems likely to become so that, in extreme but by no means rare cases, 'just' war generates 'total' war. In such circumstances, it is not some moral deficit but moral excess that accounts for the savagery in which war is conducted" (178).

Coates points to the difference between the way that German forces viewed their enemies on the Eastern Front versus the Western Front in WW II as a clear example of how the way one views one's enemy is reflected in one's conduct in war. He writes that the German's "'criminal' conduct of the war on the Eastern Front [was] both heavily indebted to the racist culture and ideology of Nazi Germany .... The rules of war were to be applied only to those who are perceived as equals" (183–4).[39] I argue, therefore, that knowledge of MEC—in particular the Symmetry Thesis—is necessary to at least foster in war's participants a recognition of the equality of combatants. Without this, combatants have maintained a historical proclivity to treat their enemy as less than equal, usually for shortsighted gains. This resulting moral zeal not only directly contradicts the intention of the in bello principles, but is also directly responsible for much of the horror in war.

The problem of self-defense and rights

I have demonstrated that there are strong consequentialist reasons for the support of both the Independence Thesis and the Symmetry Thesis. Together they provide a sufficient argument for MEC because belligerent equality results in an overall better state of affairs than a world where it did not exist. Certainly, combatants who fight for an unjust side are instrumentally supporting the unjust war, but the benefits of viewing all combatants as morally equal, I believe, outweighs these negative consequences. I will address this and other objections in greater detail later.

The defenses of MEC presented, as well as McMahan's attack, have raised issues of self-defense and those of conflicting rights. Traditionally, many discussions of just war have been in terms of rights, and theorists have often attempted to draw analogies between combat and notions of self-defense. Although useful to drawing out the discussions involving MEC, both the self-defense language and the focus on rights are problematic in themselves, and the solution that I offer—a consequentialist one—avoids these issues. Although my intent is not to offer a full critique of self-defense and rights here, I think it important to show the puzzles that these theories involve and why looking to the consequences of MEC is superior.

No self-defense problems

Benbaji's (2008c) argument draws on the self-defense literature, and while his is a good defense of MEC, a better one can be made (as I have presented). It is not his reformulation of self-defense that is problematic, rather: (1) the concept of self-defense presumes a right to life as well as a right to self-defense; (2) the nature of war does not correspond to the individual-level self-defense language; (3) while self-defense, if it is true, may apply to civilians and other non-combatants, it does not seem to apply between combatants in the same way; and, (4) if self-defense is true as well as a symmetry, then how do we account for the fact that both just and unjust combatants cannot maintain the right to self-defense?

Assume that self-defense is true. If so, then there is a puzzle of the unique context of a combatant's role that changes her right of self-defense. Ramsey (1983) writes that

> The objective of combat is the incapacitation of a combatant from doing what he is doing because he is this particular combatant in this particular war; it is not the killing of a man because he is this particular man. The later and only the later would be murder.
>
> (502)

In other words, a person, A, normally has a right not to be killed against another person, B; yet, in combat, B can kill A, and vice versa. It seems clear that no domestic analogy captures the same changed relationship found in combat, and analogies such as the police confronting a bank robber fail to capture this changed relationship. Add McMahan's (2006c) claim of asymmetry, then it seems that A, fighting for a just cause, can kill B, fighting for an unjust one, but B cannot kill A. "Troops engaged in an unjust attack," Bomann-Larsen (2007) writes, "cannot defend themselves against a counterattack without this defense becoming a part of their initial, unjust attack" (19). Thus, the ad bellum objective justice of the war defines whether B can defend herself; since her side is just, A's right to self-defense appears absolute. Again, this seems false, and it raises some troubling scenarios relating to unchecked acts on the part of the combatants fighting for a just side. While there may be legal constraints, where do the moral constraints come from?

Furthermore, war is a state of affairs between collective actors as well as between individuals, and purely individualist or collectivist views of war fall short. War is a collective endeavor from small units and crews to the state executing its monopoly on the use of force, so contra McMahan and others, war is collectivist. But, there is also a clear individualist aspect to war as well—one that some overlook in developing their accounts of war theory. Fighting a war all alone is a lonely prospect. And, it is false; it is not war. However, armies are made up of individuals—individuals who fight and die. If they survive, they bear the burdens of participating in combat. These scars may be physical and often visible: the wounds inflicted during battle. Other times, they are hidden. Certainly, explosions cause traumatic brain injury, but there is a growing concern that soldiers, individual soldiers face the potential for moral injury as well. Theory often falls short of reality, and both individualist and collectivist accounts are both partially right in their accounting of war. Wars contain both collectives and individuals.

In addition, war has no clear domestic analogy. Interestingly enough, as Walzer (1977) observes, "there are rules of war, though there are no rules of robbery (or rape or murder)" (128). Discussing combat between combatants is not the same as discussing robbery examples (nor is it the same for trolleys). Consider a different example. Self-defense theorists often point to the doctrine of double effect (DDE) to argue when a person may lose the right of self-defense.[40] And both collectivist supporters and those who advocate for an individualist view assume DDE in some form. While I think that DDE has its own internal difficulties, its problems become readily apparent when it is applied to war. Holmes (1989) observes, "double effect legitimizes every action legitimized by just necessity, provided only that one not intend to harm that he does. In fact no action whatsoever is prohibited by the principle of double effect so long as one acts from a good intention ... " (196).[41]

The proponent of DDE and self-defense must either point to the consequences of the act under consideration (as she should) or else she must introduce the complex notion of intention. Intention is of course problematic because

> Double effect allows virtually identical acts, either performed by different persons or by one person at different times, to be judged differently. Why? Because in the one case the act may be performed with a good intention, and the other with a bad.
>
> (197)

We can avoid each of these issues if we look to a consequential argument for MEC instead.

No rights issues

Closely corresponding to the concept of self-defense is the concept of rights. In war, there are not only the legal rights, which I agree seem to exist, but there also seems to be an assumed set of natural rights that inform some defenses of bellum justum, and in this project, MEC. Often, just war discussions revolve around a war-right to kill (and how that conflicts with self-defense, for example).[42] Even Walzer (1977) notes that the war convention can be best to make sense of within the language of rights (xxii).

Using rights language in describing war is problematic. For example, when war is viewed as a collective endeavor, individual rights discussions get lost in translation. As Bomann-Larsen (2007) points out, "if we are to claim that combatants lose immunity en masse, they must be regarded only in their capacity as representatives of a collective agent, not in their individual capacities. This, however, causes problems for the right-based approach, which is individualistic" (89). The onus is, therefore, on the rights proponent to bootstrap the individual rights discussion with a complex condition of war as a collective endeavor. For those who stubbornly hold to an individualist account of war, they should remember that is very rare for combat to be waged between two individuals. The battles are fought by groups of individuals: crews and units and teams and often the weapon systems are just that—systems consisting of individuals. In most cases, war is a collective endeavor and agency is more complex. Even if one presumes that a combatant's role determines her rights, the rights-based approach must account for the corresponding duties and the story of how a combatant thus loses and gains different rights at the individual level. To avoid this complex, potentially problem-laden discussion, one should argue for MEC because it results in a morally better state of affairs.

Some theorists, including McMahan, may argue that I am merely conflating moral equality with legal equality, or they might say that even if I treat the moral and legal combatantcy as distinct, I both use them interchangeably, and I assume that the legal equality is grounded in the moral equality. I will address these in reverse order. Yes, I hold that the legal combatantcy (as well as noncombatancy and the legal definition of "civilian") are grounded in morality—in particular they are grounded in moral statuses. And, while I believe that this is the case, I make no further prescriptive claim that they should or that all laws have moral grounds. Second, I do not use them interchangeably, but because they are related in a person's status in war and the words are often the same, it might appear as though I do, Finally, it should be clear to the reader that the moral equality of combatants exists as an MEC world is consequentially better than a non-MEC world, and that these legal grounds for combatant equality, the ones that both orthodox and revisionist theorists agree are necessary, which also seem prudential, actually reflect this moral equality.

Moral equality of armed contractors?

> The ambiguous relationship between governments and PMCs leaves companies open to arbitrary treatment by combatants or other countries if they stray over borders. They are combatants under the Geneva Convention if they bear arms and are clearly working on behalf of one side in the conflict, yet they could also be treated as non-combatants if they do not wear recognized uniforms or are not under military command.
>
> (Isenberg 2009, 137)

If, as Walzer (1977) observes, war is combat between combatants and MEC is true, then where do PSC contractors fit? One might take the position that PSC contractors are combatants (i.e., they are armed, etc.) but, since they often do not wear uniforms or fall under military command, they do not seem to meet the legal definition. One could claim, as the US government does, that they are civilians; yet, these contractors openly bear arms and are often involved in fighting, although primarily in self-defense. Byers (2005) writes, "Soldiers are legitimate targets during armed conflict. Killing members of the enemy's armed forces is one of the goals of military action. Still, soldiers – [i.e.,] 'combatants' – benefit from some protections, including the prohibition on the use of certain types of weapons" (127). Combatants are also granted POW protections. If PSC contractors are not combatants, then they do not have these privileges. A third choice is to ignore the armed contractor phenomenon altogether as insignificant; however, the numbers show otherwise. According to a CRS report, "As of March 2009, there were 68,197 DOD contractors in Afghanistan, compared to 52,300 uniformed personnel. Contractors made up 57% of DOD's workforce in Afghanistan"[43] and while that number decreased as operations in Iraq were concluding, there is an anticipated increase in total PMC numbers as the Syrian conflict and fight against Al Qaeda continues, and the fight against ISIS broadens. There is even a sense that a "Second Contractor War" is emerging (Hagedorn 2014). The answer remains unclear; even Erik Prince (2014) acknowledges that PSCs defy traditional legal classification. PSC contractors also defy moral classification.

I have argued that the MEC exists as it is a consequentially better state of affairs. If one holds that PSC contractors are non-combatants, then they obviously do not enjoy moral equality with combatants. However, I think that this story is incomplete. PSC contractors are not true non-combatants such as civilians, medical personnel, or chaplains. Nor are they combatants with equal moral standing to soldiers. I will briefly use the in bello principle of discrimination to help distinguish between combatants, and non-combatants, and civilians. I will then attempt to categorize PSC contractors and show that they are not legal non-combatants nor civilians, but rather a new legal, as well as moral, category of combatant. I will show that under this new category, PSC contractors, while combatants to a degree, are not morally equal to combatants or non-combatants; their moral status is unique. Furthermore, because this new category is not recognized in international law, their legal status also remains uncertain. Because of both their moral inequality and legal uncertainty, PSC contractors should therefore not be hired by states unless in extreme circumstances.

Jus in bello principle of discrimination

> Sometimes, at least, it matters to soldiers just whom they kill.
>
> (Walzer 1977, 45)

What traditionally separates the concept of murder and justified killing in war is the status of the belligerent participants. One is permitted to target combatants and not permitted to target non-combatants, but in practice the distinction is usually not so clear.[44] Although the line between combatant and non-combatant is contentious, most would agree that targeting innocent civilians is forbidden. Nagel (1979) discusses the problem of target discrimination when he notes that

> If one makes no attempt to discriminate between guerrillas and civilians, as it is impossible in an aerial attack on a small village, then one cannot regard as a mere side-effect the deaths of those in the group that one would not have bothered to kill if more selective means were available.
>
> (59)

The in bello principle of discrimination has been codified in international law. Article 48 of the Geneva Convention states:

> In order to ensure respect for and protection of the civilian population and civilian objects, the Parties to the conflict shall at all times distinguish between

civilian population and combatants and between civilian objects and military objectives and accordingly shall direct their operations only against military objectives.

As I have discussed, the in bello principle of discrimination and legal distinction appear clear-cut. Traditionally, this protection has been granted, Walzer (1977) observes, "only to those people who were not trained and prepared for war ... women and children, priests, old men, the members of neutral tribes, cities, or states, wounded or captured soldiers" (43).

While limiting civilian casualties and destruction to infrastructure is certainly a benefit of target discrimination, it also reflects

> An important facet of traditional interstate relations: members of the army are organs of the state and, when captured, are expected to benefit from this official status. Hence, like diplomats, they may not be subjected to prosecution by the capturing state for taking part in the conflict—namely killing, inflicting grievous bodily harm, carrying firearms, and so on.
>
> (Doswald-Beck 2007, 116)

In other words, when captured, combatants become POWs. Legal non-combatant status was not historically granted to civilians but was reserved for a state's official non-combatants (including captured combatants, medics, etc.).

As Doswald-Beck points out, these legal status categories do not apply in non-international conflicts, such as between states and rebel forces, nor do they apply to mercenaries and terrorists, and they may not apply to PSC contractors as well. These legal distinctions between status categories are not merely for convenience; the legal statuses reflect corresponding moral status differences—differences that are captured in the LOAC. The combatants' legal belligerent equality also includes the moral equality presumption that they are morally innocent—innocent of the decision to go to war and innocent of wrongfully killing other combatants.

Combatants, non-combatants, and civilians

Underlying the categorization in question is the notion of innocence and the state of being innocent. Although in the war context, "innocent" can be deceiving (and also seems to imply that there are guilty persons, too, on the battlefield), historically the notion that innocents cannot or should not be targeted dates back at least to Grotius (Grotius and Kelsey 1962), who offered that "no action should be attempted whereby innocent persons may be threatened with destruction." Alternatively, we could approach the problem from the other direction: one may target only these who are engaging in harm. Yet, this other approach is problematic as well. As Benbaji (2008c) notes, there are many legally targetable soldiers, for example, who are not actually engaging in harm. Consider the problem of the sleeping soldier.

Just who are the "innocent" in bellum justum? Recall that Walzer thought that women and children, priests, and so on were innocent. Anscombe (1981) wrote that the innocent in war are "all those who are not fighting and who are not engaged in supplying those who are with the means of fighting" (67). One might simply state that they are the bystanders of war.[45] Innocents in war are morally and legally immune to intentional attack, but the term "innocent" is not used in LOAC. Instead, one finds, "protected persons," "civilians," and "non-combatants" to describe the category and status of different individuals to contrast them with combatants.

According to the Geneva Convention, "civilians" are protected persons and includes all "not taking active part in the hostilities, including members of the armed forces who have laid down their arms and those placed hors de combat by sickness, wounds, detention, or any other cause ...".[46] Adopting the legal definition, one can briefly define "civilian" as:

CIV: S is a civilian (CIV) if and only if S is not taking active part in hostilities.

Alternatively, one might presume that a combatant is one who *is* taking part:

CBT: S is a combatant (CBT) if and only if S is taking active part in hostilities.

Of course, CBT is too simplistic according to international law and any underpinning moral reasoning. Certainly "taking part in hostilities" appears necessary, and includes the moral criteria of not engaging in harm, but the LOAC and other theorists have argued that "Targeting off duty, combat related military personnel, however, would not violate the principle of discrimination, since such personnel are combat related whether they are on duty or off duty, whether they are in or out of uniform, whether they are awake or asleep" (Regan 1996, 89).[47]

According to the LOAC, combatants "should be in a chain of command, wear identifiable insignia, carry their weapons openly, and act in accordance with the laws of war" (Byers 2005, 118).[48] Modifying CBT to include these legal standards, one gets:

CBT*: S is a combatant (CBT*) if and only if S

1   is taking active part in hostilities;
2   is under military control (chain of command);
3   wears uniform or other identifiable insignia; and
4   carries weapon openly.

The proviso to act in accordance with the LOAC is not what defines a combatant simpliciter; rather it helps identify those combatants who are illegal or unlawful combatants, so I will not add it here. Additionally, (2) seems to include the idea that S is a member of a state's armed forces. While this may be true, it is not always so. Military medical personnel and military religious personnel, while they are members of the armed forces, and so on, they are not considered combatants by the LOAC, and are thus not combatants, CBT*. CBT* are authorized to use force, and they are legitimate targets. Civilians, on the other hand, may not legally use force (except in self-defense), nor can they be targeted.

Interestingly, as discussed earlier, McMahan (2008) argues that there is no moral distinction between combatants and non-combatants; thus, there are times when it may be morally required to target non-combatants.[49] Nevertheless, what inspires him to make this argument is the empirical evidence that civilians are often culpable, not only culpable for the decision to go to war, but they are also significantly supporting the war. Examples could include camp followers, scientists who design lethal technologies, the often-discussed munitions factory worker, the taxpaying citizen, or the contractor who mans the sophisticated weapon system. Some of these, it seems, ought to be targetable, while others should not be. The question, then, is where to draw the line.[50] I think that answer remains unclear. For now, however, I will call those civilians who are significantly supporting the war as "Materially engaged in the war effort," or CIV_{WE}. The idea here is to establish a category where certain civilians, by their actions (or perhaps even their role) become legitimate targets under discrimination. It is not as important at this point to decide who falls into this addendum to CIV; rather, it is well enough that one acknowledges the possibility that a person might be a CIV_{WE}.[51]

Nonetheless, combatantcy comes with an added privilege that civilians do not enjoy. If captured during war, combatants gain prisoner status (POW); no longer combatants, they are now not targetable. Civilians are usually not considered POW-eligible. Legally, civilians may be detained for criminal activity, but according to the LOAC, because they were not taking an active part in the hostilities, they would normally not be entitled to become POWs if captured. (Of course, there is an exception I will return to shortly.) In fact, since civilians are not under military control, not in uniform, etc., if they are caught participating in combat, they may be held as illegal combatants. Alternatively, civilians who are not members of the armed forces may attain combatant status if, for example, they become members of police forces, and so on, just as one could join up tomorrow. CBT* status or CIV status may change. Furthermore, it is unclear is whether CIV_{WE} are to be given POW status, or are they also potential illegal combatants?

The Geneva Convention does however attempt to clarify their status. Article 4.A (4) states:

> Persons who accompany the armed forces without actually being members thereof, such as civilian members of military aircraft crews, war correspondents, supply contractors, members of labor units or of services responsible for the welfare of the armed forces, provided that they have received authorization from the armed forces which they accompany, who shall provide them for that purpose with an identity card.

Although it seems strange to say that an ID card can determine one's status, this is a common procedure among the PMCs hired by the US government. However, this also raises several issues. First, not all PSC contractors have the ID card granting them civilian status, nor are they necessarily accompanying the military (as the ones hired by the State Department). Second, even if they did, there is some uncertainty with the phrase "taking active part in hostilities" that appears problematic. If A, normally a CIV, does take part in hostilities, then she is either an illegal combatant or is now a CBT* due to a change in her status. If she is not an active participant, as most PSC proponents claim, then she either remains a CIV or is a CIV_{WE}. How does one determine whether A is an active participant?

Where do armed contractors fit?

In wartime, professional soldiers are combatants, and I have discussed the legal and moral equality of these combatants. The question is whether the PSC contractor is a combatant as well. The US government of course maintains that contractors, including armed PSC contractors, are CIVs (Elsea, Schwartz, and Nakamura 2008; Rothwell 2004; Doswald-Beck 2007, 128). The Montreux Document also seems to reinforce their Status as CIVs. It says that PSC personnel:

> b) are protected as civilians under international humanitarian law, unless they are incorporated into the regular armed forces of a State or are members of organized armed forces, groups or units under a command responsible to the State; or otherwise lose their protection as determined by international humanitarian law.
>
> (Maurer 2008, 11)

However, CIVs are protected persons and "are neither members of the armed forces of a belligerent (that is to say, a party to the conflict), nor do they play a direct or active part in the hostilities" (Byers 2005, 118). Yet, even if one accepts that PSC contractors do not take active part in hostilities, it seems clear that they are armed, and their services, which include providing on-site and mobile security, protection of locations or persons, and the security escort of supplies and personnel, are military related. While these are perhaps not key military competencies, the services PSC s provide do raise several concerns: (1) military-based skills are required, and PSC contractors are recruited and hired for having these skills; (2) these missions seem to fall into the "materially support the war effort" category; (3) if the PSC contractor did not perform these tasks, then military personnel would have to do them; and, (4) as Byers (2005) notes, the "extended involvement of these contractors and activities traditionally reserved for military personnel is obfuscating the all-important distinction between combatants and civilians, with potentially serious consequences" (118–19). Labeling PSC contractors as CIV does, however, allow states a way of reducing the reported numbers of military personnel deployed and provides an implicit banning of their intentional targeting. Nevertheless, PSC contractors are not CIVs, and although current home-state law suggests they are, they should not be. Perhaps they are CIV_{WE}.

Determining whether PSC contractors are CIV_{WE} or combatants thus must be based upon whether they "take an active part" in war. One way to view "taking an active part" is to simply state that:

> S "takes an active part in hostilities" if and only if S is actually fighting or is in combat.

However, this conception immediately becomes troublesome because although military personnel are combatants, not all are engaged in combat; many will never actually be shot at nor attack another. There is a kind of potentiality at work here; her role of a soldier, that she is a soldier, makes her a combatant. Armed contractors have also been and will be involved in actual fighting. For example, consider the Fallujah incident where four Blackwater contractors were killed and mutilated. Isenberg (2009) observes, "They carried weapons, but unlike soldiers, they presumably had orders only to use them in self-defense within the scope of their contract and the coalition's rules of engagement. In addition, they had a quasi-military mission to defend convoys from Iraqi insurgents" (77). Unfortunately, this kind of combat confrontation has frequently occurred. I have discussed the Blackwater employees who were fighting to defend the CPA complex in Najaf, "even relying on their own helicopter to resupply the forces and ferry out a wounded Marine" (Avant 2005, 233; 239),[52] and HART Group personnel were involved in a firefight in Kut the next day. There is an old saying, "If they are shooting at you, it's combat." In addition, US Army combat awards designate eligible individuals as having been personally present and under hostile fire; therefore, if they were soldiers, then the contractors discussed might have received combat badges. They did not.

However, it must be said that PSC contractors are neither intended for nor hired for combat operations, and the contracts are written to avoid this legality. Nevertheless, these same contracts discuss the risks, including direct and indirect fire, which the contractor may face. Outfitted with armor protection, individual and squad level weapons, and even including helicopter support, PSC contractors are surely prepared to, if necessary, receive hostile fire and return their own. Legally, at least for US PSC contractors, the Defense Federal Acquisition Regulation Supplement states that "contractor personnel are not authorized to use deadly force against enemy armed forces other than in self-defense" or "when necessary to execute their security missions to protect assets/persons, consistent with the mission statement contained in their contract" (quoted in Krahmann 2007, 109).[53] Therefore, while one might say that PSC contractors have and do fight—that is they do see combat—this fighting is, according to current law, only defensive in nature. We might reformulate "taking part in hostilities" to read:

> S "takes an active part in hostilities" if and only if S is (or potentially is) conducting offensive operations.

(Here, for simplicity, I use the US Army's definition of offensive operations. Offensive operations mean "attack simultaneously throughout the area of operations (AO) to throw enemies off balance, overwhelm their capabilities, disrupt their defenses, and ensure their defeat or destruction" (2008, 7.2).)[54] Furthermore, this formulation also captures how PSCs avoid anti-mercenary legal action, as international law labels as mercenaries those persons who "take a direct part in the hostilities," but are not members "of the armed forces of a Party to the conflict" (Geneva Conventions Protocol I, Article 47(2)(b) and (e)).[55] Thus, PSC contractors, while not CIVs, may be CIV_WE.

True, PSC contractors are not currently hired to conduct offensive operations; however, the lines between offensive operations and defensive ones are often blurred, especially in counterinsurgencies where we find many of the PSCs being employed. In the military, one may speak of an operation being an offense one, but depending on the level of engagement and unit levels involved, some subordinate units may be on the offense while others are conducting the defense to support the overall defensive plan. For example, a counterattack in a defense or assaulting one's way out of an ambush are both offensive actions taken to allow the defensive plan to succeed or to survive an attack. Therefore, even PSCs hired for defensive efforts may find themselves, at the individual or small group level, on the offense.

When PSCs are providing security or policing services, they may be authorized to use force necessary to defend themselves, other persons, or property. However, as Isenberg (2009) notes, "given the fluid nature of the current security situation in Iraq, it is sometimes difficult to discern whether civilian security guards are performing law-enforcement duties or are engaged in combat" (155). Not only are PSC contractors susceptible to find themselves actually fighting, but the services they provide combined with an uncertain environment often make it indeterminate whether the combat is offensive or defensive. Even during innocuous services such as supply escort, combat is almost inevitable; to deny otherwise is wishful thinking. Verkuil (2007) points out that

> Offensive and defensive may be a distinction without a difference. Keeping operations defensive in a setting where there are no front line combat zones is a virtually impossible task. Once a decision is made to permit the private military to enter the battlefield, combat support services evolve into combat services. Thus, to an increasing extent, the private military in settings such as Iraq are part of the combat mission.
>
> (27)

And, what should we say of cases of hot pursuit or when preventative, offensive measures are used? Since PSC contractors operate in harm's way, and they are equipped to do so, and it is very likely that they will see combat, it seems unfounded to continue to call them CIV_{WE}.

Here are some examples. During Desert Storm, Vinnell contractors working with the Saudi National Guard deployed with their Saudi units into combat at Khafji.[56] A California Microwave Systems plane carrying four contractors was shot down in Colombia over FARC territory. Eight DynCorps contractors "accompanied Iraqi police who raided the Baghdad home and offices of former US favored Iraqi politician Ahmed Chalabi on May 20 [2004]," and DynCorps, providing assistance to counter-drug operations in Columbia, was allegedly in combat encounters with rebels on the ground as late as 2001. When the Program Management Office in Iraq solicited bids for a comprehensive security contract, they requested up to 75 teams to conduct tasks, including "mobile vehicle warfare" and counter sniping, with the ability to "protect sites against indirect fire and attacks by small units."[57] While perhaps isolated incidents, they point to PSCs operating more as CBT* than as CIV_{WE}.

Some may argue, however, that these incidents merely show that the defensive tasks PSCs conduct do include a risk of combat. I agree, but not only is it a very high risk, I also think that this small sample demonstrates that a PSC's line of work entails combat. Furthermore, there are other areas where contractors are being used that definitely appear to cross the offensive line. Information war has both offensive and defensive sides, for example, and contractors are hired to conduct or support offensive information warfare. The same can be said about psychological operations. Isenberg (2009) observes, "Even the US Special Operations Command, the umbrella organization for all U.S. special operations forces, has sought the help of private contractors to make slick multilingual audio, video, print and Web packages to support its global psychological war against terrorism" (25). Furthermore, contractors also fly and maintain UAVs. If used for offensive operations, perhaps by flying armed UAVs, then it could be argued that the UAV contractors are combatants. Even if they are not CBT*, they might create the impression they are, and this is just as problematic.

The in bello principle of discrimination, to be effective, entails that one can discern between combatants and non-combatants. Although this is genuinely difficult in practice, as Navy Seal Marcus Luttrell (Luttrell and Robinson 2007) recounts from Afghanistan, the introduction of armed PSC contractors on the battlefield increases this risk. On the one hand, contractors are sometimes in uniform, ones similar to what the participating soldiers wear, and at other times they are in civilian clothing.[58] Furthermore, PSC contractors are armed in much the same way as combatants, even using helicopters and other military-type vehicles. As Singer (2003) writes, "when a local guerrilla force is hit by US-made gunships, piloted by US citizens, and using US tactics, is completely logical for them to assume that the attack involved official United States forces" (212). I worked with many different contractors in multiple deployments, and it seemed as if each company had its own rules regarding "uniforms." Some wore civilian clothes; others wore what we colloquially called the "contractor uniform" or "Blackwater uniform" of khakis with many pockets. Some wore uniforms with different insignia. With so many options, it was often confusing for our soldiers. This inability to distinguish between combatants and PSC employees, also blurs the lines for others, non-PSC personnel such as other contractors or even journalists, who are often either mistaken for combatants or grouped with them in the eyes of the enemy. Isenberg (2009) relates the story of an Iraqi interpreter, who said, "If the insurgents catch us, they will cut off our heads because the imams say we were spies" (45).[59] Although the distinction between combatants and PSC contractors may be readily apparent to some, it is not often the case. Furthermore, if the services PSCs provide amounts to, for all intents and purposes, combat, then "they become lawful targets for lawful enemy forces during the fighting, and, if captured by such forces or an enemy government . . . , they could potentially be prosecuted criminals for their hostile acts" (155).

Are PSC contractors morally equal?

As I stated, according to international law, PSC contractors are not combatants, CBT*. However, clearly they are not CIV nor CIV$_{WE}$ either. I propose that there be a derivation of CBT*, call it CBT$_{PSC}$, which accounts for the problematic categorization of PSC contractors. Although CBT$_{PSC}$ is not recognized by the LOAC, I think it should be. By being armed, conducting military related services, and operating in an uncertain wartime environment, and precisely because, even if they will not intentionally engage in combat, they are there, armed and ready, on the presumption that there will be combat, PSC contractors are combatants of a certain sort. Further, one must add the inherent problem of indeterminate discrimination. It is only the fact that they are not members of the military or under military control[60] that they are not CBT*; they are instead CBT$_{PSC}$.

Assuming that MEC is true, CBT$_{PSC}$ are not the belligerent moral equal of CBT*. CBT* are equally allowed to be targeted by other CBT*, and they are permitted to kill other CBT*. CBT$_{PSC}$—while perhaps targetable like CIV$_{WE}$ can be—still are not granted license to kill other combatants, CBT* or CBT$_{PSC}$. They may only use force in defense of persons or property. There is thus a moral asymmetry between PSC contractors and combatants. Neither are they necessarily just or unjust; they are combatants of a different kind and are not morally equal under jus in bello. It is this inequality that captures the uneasiness that opponents to the hiring of PSCs profess to have.

Some objections

> But by and large we don't blame a soldier, even a general, who fights for his own government. He's not a member of a robber band, all willing wrongdoer, but a loyal and obedient subject and citizen, acting sometimes at great personal risk in a way he thinks is right.
>
> (Walzer 1977, 39)

I have argued that MEC is true and forms an integral part of bellum justum. I have also demonstrated that PSC contractors are neither civilians, as is legally assumed, nor are they morally innocent. Rather they are a kind of combatant. However, a PSC contractor's combatancy is not the same as a soldier's; they are not morally equal on the battlefield. While I have discussed several objections along the way, three remain that I believe need more attention.

McMahan would counter that my consequentialist argument for MEC is similar to his acknowledgment that a legal conception of belligerent equality is consequentially better, but a moral equality between combatants is impossible. Second, some may argue that I am trying to shift to a consequentialist view of just war, which many might contend is a non-starter. Finally, I want to briefly comment on other kinds of soldiers—such as conscripts, children, and so on, and why they are morally equal to professional soldiers while PSC contractors are not.

Legal versus moral equality of combatants

McMahan might accept part of my argument. He would agree with my consequentialist reasoning that it would be better to have belligerent equality in part to limiting the destructiveness of war. For McMahan (2008), however, it is a legal belligerent equality, not a moral one. Therefore he concludes, "that the only feasible option, at least at present, is to grant legal permission to both just and unjust combatants to fight and kill in war" (28).

Nevertheless, contra McMahan, it is the very consequentialist reasons he relies upon for adopting a legal belligerent equality, I contend, that also provides the moral grounds as well, and this is why MEC is true. McMahan is of course not a consequentialist, but by asserting that there is good reason to adopt a legal version, he could also apply the same logic for MEC. Even his just-unjust distinction relies upon consequences grounds—all acts by unjust combatants are instrumentally supporting the unjust cause. This may be true; however, the costs of war without MEC, the principle that also informs the legal equality of combatants, are graver than the instrumental harm of the unjust side, especially when combined with the consequences I outlined earlier. Walzer (1977) writes, "Without the equal right to kill, war as a rule-governed activity would disappear" (41); this should be reason enough why MEC exists. These rules are grounded upon the moral relations in war, including MEC. There is not only a need for the legal belligerent equality; a MEC exists.

My solutions entail adopting a consequentialist approach to war

Some might contend that I have not only provided consequentialist arguments for MEC and a belligerent inequality of PSC contractors, but that I am also trying to advocate a consequentialist approach to bellum justum. Theorists, such as McMahan (2008, 24), Walzer (1977, 129), Rodin (2007, 597), Hurka (2008), Benbaji (2008b, 5) and Bomann-Larsen (2007, 67), have acknowledged that one should consider the consequences of war and that there are some consequentialist reasons for adopting just war principles. However, bellum justum has been traditionally viewed as either a theory of rights or the deontological limitations that prevent total, unrestricted war, which could result from purely military consequentialist calculations.

Walzer (1977) writes, "The war commission invites soldiers to calculate costs and benefits only up to a point, and at that point it establishes a series of clear-cut rules—moral fortifications, so to speak, that can be stormed only at great moral cost" (130). I agree. Furthermore, I have argued elsewhere for a sophisticated utilitarian approach to jus in bello (2009). Nevertheless, arguing for a consequentialist approach is beyond the scope of this project and is not needed for my argument. I suggest that there are good consequentialist reasons to counter McMahan's argument and defend the MEC. Furthermore, assuming MEC, PSC contractors are not the moral equal of soldiers, nor are they civilians in the legal sense.

One might imply that I am adopting a rule utilitarian argument—that is., that MEC is the rule that results in the greatest utility. While I do agree that MEC results in greater utility, here I am not wedded to utility, nor am I advocating a Brandt-like rule utilitarian approach to just war (1972). If MEC is true on consequentialist grounds, then it may require us to rethink a deontological approach to just war; however,

I think that it is clear that acknowledging MEC, and conforming to its theses, restricts the brutality in war. Moreover, it helps one understand why the hiring of PSCs is morally problematic.

If there is no moral difference between child soldiers, conscripts, and professional soldiers, then why are PSC contractors any different?

Some may point out that while I spend a great amount of time arguing for MEC, I focus on volunteer and professional soldiers only. They might ask why I have ignored children soldiers and conscripts in my schema, for these cases are also morally troublesome. I agree that, especially in the use of children as soldiers, there is something definitely upsetting at work. Children soldiers, abducted from their families and villages, often drugged too, are indoctrinated in killing. They may be beyond help after war's end, but this child soldier, once she takes up arms and becomes a threat, nonetheless is a combatant, the same as soldiers who are con-scripted or those who make a career from soldiery.[61] Both child soldiers and con-scripted ones are coerced to fight. One might further say that they are victims of their predicament, and to some, the gladiator analogy appears appropriate.

Nonetheless, the Independence Thesis stands, as these kinds of soldiers may not even have access to ad bellum information, and they are in little position to con-template the political situation, let alone their own. Also, regrettably, the Symmetry Thesis holds as well. Because they take up arms as combatants, they are targetable, and like other combatants, they can be killed in the process. Children soldiers, while perhaps unusual combatants, are not illegal nor immoral ones (unless they commit war crimes). In addition, special consideration must be made at conflicts' end to help reintegrate them into society. However, these issues do not make them less than combatants—less than morally equal. PSC contractors are different because they morally and legally cannot attack other combatants, $CBT_{PSC,}$ or CIV_{WE}. They can only use force in defense of persons or property.

Therefore, although I consider the discussion of children soldiers as a very important one, their moral status as combatants does not affect my argument. MEC limits war's brutal nature. Moreover, the employment of PSC contractors may not only be legally challenging, but also raises moral concerns because they are neither the legal nor moral belligerent equals of soldiers.

The moral and legal status of PSCs

McMahan (2006c) offers some sage advice; he writes, "We ought ... to encourage moral reflection, even among active duty military personnel, and not only about what it's permissible to do during war but also about when it's permissible to par-ticipate in war at all" (392). Nevertheless, accepting these words of wisdom does not entail that MEC is false. In this chapter I have sought to show that while McMahon is correct in highlighting the weaknesses of the conventional, orthodox position of MEC, rejection of MEC is clearly an unacceptable mistake. It is the Independence

Thesis that allows us to critically evaluate conduct in war, and Symmetry Thesis helps us discern the distinction between war's participants—from combatants to contractors to children—and it provides guidance for target discrimination, thereby providing some limitations on war's brutal nature.

Perhaps, as some point out, these are pragmatic concerns. Nevertheless, it is these very pragmatic reasons that consequentially form the moral grounding for MEC. Other orthodox view supporters must wade through the problems of liability, self-defense, and intentionality to argue for a weakened version of MEC at best. My solution avoids these problematic strategies, clearly pointing to why MEC is true. Furthermore, PSCs clearly are neither the moral equals nor the legal equals of combatants; PSCs are also not civilians or even CIV_{WE}. PSCs form their own status, CBT_{PSC}, which is why they are both morally and legally problematic.[62] With this framework and the CBT_{PSC} status, we can better discuss the Najaf battle and how the Blackwater contractors were legally and morally distinct from the marines, but they were within the law and morally permitted to defend the CPA site and each other.

Notes

1 Also see Scahill (2007, 123).
2 See Walzer's (1977) Chapter 3 for a widely accepted account of the traditional (sometimes called the orthodox) view of MEC.
3 Also see, McMahan's "Collectivist Defenses of the Moral Equality of Combatants" (2007) and "The Morality of War and the Law of War" (2008).
4 See Holmes (1989, 87–95), Regan (1996, 104); Hurka (2008, 135–6), and Bomann-Larsen (2007, 39–40; 78–89). Note another difficulty with discrimination: the wearing of uniforms. The Geneva Convention states that combatants must wear uniforms; it then would conceivably make it easier to distinguish between combatants and non-combatants. So, what about guerrillas who do not wear uniforms or private contractors who wear uniforms? I will discuss this problem later in the chapter.
5 This has been subject of much recent debate. See Rodin (2007), Ceulemans (2008), Schoonhoven (2003), Zupan (2006), Pattison (2008), Sterling (2004), Miller (2004), Shalom (2011), Rocheleau (2010), and Wertheimer (2007).
6 In particular there are two recent proposals by Yitzhak Benbaji and Lene Bomann-Larsen. See Benbaji's "A Defense of the Traditional War Convention" (2008c) and Bomann-Larsen's *Reconstructing the Moral Equality of Soldiers* (2007), respectively.
7 McMahan raises three objections to the boxer analogy. First, McMahan does not think it is possible for all combatants in all wars to consent to being killed (381). Second, McMahan suggests that even if the combatants did consent (i.e. a duel), it still would not be permissible to kill the other. Lastly, McMahan argues that even if all combatants consented to being killed by their opponents, that wrongdoing is still being committed. However, McMahan is too hasty here. Consent, especially mutual consent is truly difficult to measure. Furthermore, by being soldiers, fulfilling a role per se, seems to impart a least a weak, tacit consent. Benbaji (2008c) recognizes this apparent consent based upon the general acceptance of rules regarding the conduct of war, such as jus in bello, rules Van Creveld (2008) notes have been practiced since ancient times in both primitive and modern cultures.
8 Gladiators have not consented to fight one another in the same way as boxers. They are coerced. Yet, McMahan (2006c) argues that the threats used to coerce combatants also cannot justify their killing one another. Although he acknowledges that the coercion can

be severe enough to exculpate unjust combatants, their killing of just combatants would still not be permissible. Yet, as Schoonhoven (2003, footnote 14) points out, "There may be a confluence here with the other major consideration that is often adduced in support of [MEC], viz., the power of the state to coerce or compel soldiers to fight ... and while states are not the only entities with the power to compel, their ability in this regard is considerably greater." While McMahan would counter that this fact merely exculpates the unjust combatant, some theorists, such as Bomann-Larsen (2007; 2004), have argued that exculpation is sufficient for an equal right to kill. I agree with McMahan, though, that Walzer's analogies are inadequate for MEC.

9 As Benbaji (2008b; 2008c) noted, it seems reasonable for combatants, since they cannot determine whether their side will be just, to consent to rules that govern conduct in war. McMahan (2006c) thinks is false, "since potential combatants would have more reason to accept a principle that would require them to attempt to determine whether their cause would be just." This view, he contends, may result in fewer wars fought for unjust causes; combatants would be "less likely to be used as an instrument of injustice" (384). Nevertheless, entering this conceptual war convention does have other benefits, including post bellum immunity for all soldiers. We consider that soldiers are legally equal if they "possess the same set of rights and duties" (Benbaji 2008c, 8), including in bello duties. One might consider this a Rawlsian war convention, where self-interested agents deliberate on what the rules of war should be (Rodin and Shue 2008, 9). What rules of war would they accept? I believe their rules would include MEC.

10 Combatants not only have a tacit acceptance of MEC but MEC also captures their commitment to and perceived obligations from their military institutions and home states. As even McMahan (2006c) observes, militaries "depends on a division of moral labor" and are "organized hierarchically with a rigid chain of command" (384). This institutional commitment reflects a partial surrendering of some individual autonomy (concerning ad bellum decisions) that not only makes a military efficient, but also highlights the Independence Thesis. This reasoning is not new in war; Vitoria (Vitoria et al. 1991) noted: if soldiers "fail to obey the prince in war from scruples of doubt, they run the risk of betraying the Commonwealth into the hands of the enemy" (311–12).

McMahan (2006c) disagrees. First, an institutional commitment cannot apply universally, refering to Nazi soldiers and the Iraqi Republican Guard. Nevertheless, even though institutional commitments can still result in unjust combatantcy, it does not follow that it must. We are more critical of the Nazi soldiers because of their cruelty and acts of war crimes than their acting in unjust wars. Being used as an instrument of the state's unjust policies points to a soldier's shared victimhood instead. Both sides are public agents, responsible for securing their communities—their institutional commitments.

Some might argue, using a domestic analogy, that an executioner is morally absolved from determining whether the one she executes is guilty or not; the decision of guilt is outside of the executioner's purview. States do not provide, counters McMahan, the "impartial and epistemically well-grounded decision" (388) analogous to our courts system. But, especially in democracies, this seems false. In the US, for example, it is not just the Congress that can declare war or the President as Commander-in-Chief who can deploy forces. Both branches of government are subject to checks and balances of the judicial branch and the people. Certainly, both the government and its citizens can launch an unjust war, but so too can a judge and jury of peers condemn an innocent man.

11 The epistemological argument concerns the combatant's not knowing whether the side she fights for is just. Part of this conception historically dates to when soldiers seemed more as pawns in the wargames of the princes (i.e. invincible ignorance), and as such is an artifact that should, some claim, be discarded. Like the aforementioned executioner, however, the soldier is considered outside of the ad bellum decision. While McMahan (2006c) acknowledges that unjust combatants believe their cause is just, it is not reasonable to assume that one's cause is just. Thus, the claim involved in the epistemological

argument seems to point to the degree of confidence a combatant has in her belief that the war is (un)just. To what degree then would be satisfactory to presume a combatant could make reasonable assessments of the ad bellum decision? There are two schools of thought in the epistemological debate. McMahan argues that unless one has a high degree of epistemic certainty that a war is just, she ought to abstain. On the other hand, proponents of MEC, such as Zupan (2008), argue that one should follow her institutional commitments, unless she has a high degree of epistemic certainty that the war is unjust. As Benbaji (2008c) notes, in either school, "epistemic indeterminacy [seems to matter] for morality" (485). Looking back to the executioner analogy, it is not the case that combatants should not consider ad bellum issues; rather, ad bellum decisions are not ones combatants make. Furthermore, since most combatants believe their war is just (and it would be difficult to prove otherwise), then the presumption is that she should honor her institutional commitments. She might be wrong; of course.

12 As Bomann-Larsen (2004) notes, "Those responsible for an unnecessary and unjustifiable war are no less responsible for the deaths of their own soldiers than for the death of the other party's soldiers. We cannot, to be sure, hold [just] soldiers accountable for the deaths of [unjust ones], but we can and do hold the political leaderships ... who committed the crimes of aggression in the first place responsible for these deaths" (148).

13 Rodin (2007) explains this well. He writes, "[Future combatants] in the original position can therefore know a priori that, at most, fifty percent of all wars (understood as the prosecution of a war by one party) can be just. If all wars are just on one side and unjust on the other side, then the percentage of just wars will be fifty percent; if some wars are unjust on both sides then the percentage will be less than fifty percent. Therefore, if at least one conflict in the universe of possible wars is fought unjustly on both sides, then the majority of all possible wars are fought unjustly, and the majority of combatants across the total class of wars will be unjust combatants" (602).

14 Perhaps there is an existential threat to what Orend (2013) considers a minimally just state (37–42).

15 Benbaji (2008c) attacks McMahan's arguments (what he calls the "purist view") from two standpoints. First, he argues for the Independence Thesis, and the historical precedents keeping in bello separate from ad bellum (2008b, 3). His second attack argues for the Symmetry Thesis based upon a reformulation of self-defense. The self-defense (SD) framework is modified, so he contends, to more closely resemble the war context from a Lockean perspective. In war, Benbaji (2008c) argues, because there is a presumption that a soldier will participate in war, she poses a threat. (Whether Benbaji holds that the threat she poses is a standing one, or a potential one, or even a temporal (e.g., future) one is unclear.) Nevertheless, her "threat posing" status is not attributed to her threatening acts of war. It is part of her role as a soldier. Benbaji writes, "Indeed, innocent, not quite innocent, and guilty enemy soldiers are all present in the battlefield" (468). Combatants are therefore mutually innocent threats. Benbaji's SD may illuminate solutions to the falling man and other innocent threat problems. I am not entirely convinced, however, that one can apply the same SD paradigm to war. Discussions of combatants' mutual right to SD does not seem to apply in war as it does in cases of falling men, trolleys, or robbery examples. If we accept this framework, then proponents of SD must (1) argue how a soldier, merely by being a member of a class and not in terms of being a personal, direct threat to his enemy, can still be killed, and (2) provide an explanation for why non-responsible, material threats may be justifiably killed (Bomann-Larsen 2007, 91–2). Benbaji's (2008c) SD reformulation, entails a lesser harm principle. For example, X is justified in killing A in self-defense, if the "lesser evil" principle recommends it (475). Benbaji notes that the amount of harm affects whether X is justified in killing A—clearly consequentialist considerations. Elaborating on a consequentialist nature of the lesser harm principle may provide a way to strengthen Benbaji's SD proposal. SD may be less

a conception of rights but rather a weighing of the harm committed. (Unfortunately, this discussion will have to wait for a later time.)

16 Bomann-Larsen considers the MEC question from innocent threat perspective, but she departs from his SD model because, unlike the fat man and his victim, combatants are morally autonomous and responsible for their actions. Nevertheless, she believes that they are not blameworthy. While Benbaji may contend that combatants are not blameworthy (in terms of ad bellum) because of their role as soldiers alone, Bomann-Larsen (2007) contends that it is a "moral conflict" that arises from both their apparent conflicting obligations being soldiers, as well as the uncertainty in the ad bellum realm (134–5). Combatants are "exculpated aggressors," and a combatant's killing in war is exculpated by her role responsibility as a soldier. Thus, "soldiers are wronged in their mutual fighting, but that they do not act wrongly by killing each other." She continues, "Their war-right to kill is a weak permission, based on mutual exculpation for fighting" (13). This combatant role responsibility drives not only the uncertainty surrounding conflicting obligations, but also separates the individual combatant from her objective, unjust cause. The independence between the collective and individual levels of war, Bomann-Larsen contends, that "makes it possible for us to separate our judgments of the collectively wrongful proceeding of an unjust war" and the moral standing of the soldiers (132). An unjust combatant, therefore, can be exculpated for her blameless wrongdoing, abiding by her institutional commitment to the state and through subjecting herself to its political decisions.

17 The institutional commitments solution lies in part on collective morality and the separation between individual and collective responsibility (Zupan 2008; Benbaji 2008b). McMahan (2007) labels this the collectivist defense. The collectivist argument surmises that one ought to view war as a collective endeavor; combatants therefore should be viewed as agents of a collective rather than as individuals. Whether one holds to a division of moral labor between the collective and the individual or one holds to a role responsibility view, either make a collectivist defense. One may be morally obligated to vote against the war yet not be obligated to refuse to fight. Walzer (1977) appears to hold this view; he writes, "[combatants] vote as individuals, each one deciding for himself, but they fight as members of the political community, the collective decision has already been made" (supra note 299). The collectivist argument can be derived from institutional commitments, and one could present a strong version of the collectivist argument (like Zupan 2008) that advocates that the agent who is responsible for the killing is the state, not the individual combatant. The institutional commitments of the combatant obligate her, like the executioner, to carry out the sentence. This division of moral labor seems to closely match the ad bellum—in bello separation. As Benbaji (2008b) notes, "Soldiers have a right not to be guided by ad bellum considerations because statesmen are under a duty to do so" (25). McMahan (2007), of course, disagrees; he asserts that (1) this obedience to the state cannot be absolute, and (2) if the unjust combatant is an innocent individually, she becomes a legitimate target as part of the collective. Nevertheless, even if a combatant's moral status is not wholly derived from her role or station (she of course retains some moral autonomy, and we do not apply a blanket guilt by association), it does not follow that her institutional commitments do not affect her moral standing. We want to be able to hold individual combatants morally accountable for their acts in war. Furthermore, a corporate conception of moral autonomy is complex at best, perhaps even impossible. As Holmes (1989) notes, "those who wield power rarely effect any of its consequences directly. They personally collect no taxes, make no arrests, and fight no wars. The power is transmitted to the wills of thousands, and sometimes millions, of other persons" (207). Holding certain political leaders or high-level military leaders accountable makes sense; corresponding punishment of culpable whole polities or individual combatants because of their collective roles does not.

18 While here Walzer is attacking Sidgwick's conception of utilitarian war, he misjudges Sidgwick as one who advocates subjective utilitarianism. What Sidgwick would actually argue for, I believe, is a sophisticated utilitarian approach to war. See my "Jus in Bello and the Sophisticated Utilitarian" (2009).

19 Cuelman (2008, 107) notes: "It is often very difficult, if not impossible, to objectively assess which side justice favors."

20 Some also include right intention as an additional ad bellum tenet. Cf. Orend (2013).

21 Orend further discusses the issue of multiple or plural causes (2013, 49–52).

22 Zupan (2008) argues that "Being under orders, trusting in his superiors, focusing on the mission at hand are such a part of the ordinary experience of being a soldier that 'knowing' his war to be unjust turns out to be something he literally cannot do" (218). The experiences Zupan describes are accurate as I have discovered speaking with soldiers of different ranks throughout the years, and these experiences are shared by many different military members worldwide.

23 As Bomann-Larsen (2004) observes, "The military is a legally, socially and politically legitimate institution, and its legitimacy is cast in terms of its service to the state and society. Soldiers represent, and act on behalf of, the state; literally being its arms . . . . It is implied that had he acted on his own, the soldier would have no right to kill" (143).

24 Benbaji (2008b) also adds, "For it is in the interest of individuals who live in a world of sovereign and mutually suspicious states to be protected by a decent state which controls an obedient army. That is, soldiers (and their families, friends and fellow citizens) benefit from the in bello agreement because it makes national defence more efficient" (8–9).

25 Rodin and Shue (2008b) reply, "if soldiers were under an obligation to initiate private military action whenever they believe this to be justified, this would seem to be a recipe for global chaos" (13). Also see (Ryan 2008). Ryan uses the examples of the filibusters in early US history to highlight the problem of private forces.

26 Christopher (2004) notes, "Standing armies are recruited, trained, and maintained to defend the community in times of danger. And although defending one's community or family may be the right of every citizen in primitive society, in civilized society that right has been taken away from the population at large and isolated in the office of the soldier. It is, therefore, a duty for the soldiers to defend those constituents who have entered a pact with them to this end" (89). Also see Hartle (2004, 28), and Walzer (1977, 39).

27 Benbaji (2008b) argues that only states that have obedient militaries can be effective deterrents (11). While I disagree with his idea that, "under a regime that obliges soldiers to ascertain that the cause with which they fight is just, armies won't be able to efficiently fight a just war" (2008c, 490), I do concur with the theory that the existence of obedient and capable militaries "might contribute to the preservation of a peaceful stable balance of power" (2008b, 11) and be an integral part of a state's defense, preventing attacks by unjust states. Benbaji's argument, though, seems to beg the question of a military institution's legitimacy. Nevertheless, we generally consider the military, writes Bomann-Larsen (2007), as "an institution of law enforcement at the international level, [which] serves the community and community values. Generically speaking, the military is established for legitimate ends (even if it does not always serve legitimate ends)" (168).

28 Also see Kutz (2008).

29 Tony Coady (2008) writes, "it is understandable that there should be a presumption that warriors are entitled to direct lethal force against opposing warriors where they have some plausible warrant for seeing them as wrongdoers or attackers, without there being any such case for attacking non-combatants" (164).

30 Bomann-Larsen (2007) continues, "Precisely because it is possible for just combatants to be war criminals, it is possible for unjust combatants to cause harm with justification" (126). See also Benbaji (2008c, 490). He writes, "Of course, soldiers have not relinquished all autonomy. Quite to the contrary. They are entitled to disobey orders that manifestly violate the rules of conduct in war; they, in fact, ought to do so."

31 Such as noncombatant immunity, necessity, and proportionality.

32 McMahan (2008) writes, "The rejection of the legal equality of combatants would also carry the risk that, in cases in which it is fairly obvious to unbiased observers that one side in a war is in the wrong, and in which the world begins to point an accusing finger at the combatants on that side and to charge them with criminal action, this will provoke defiance and a renunciation of all restraint" (30).

33 Walzer (1977) recognizes the importance of the ex ante agreement, noting that merely asserting some moral principles regarding war is insufficient. Rather, it takes "the work of men and women (with more principles in mind) adapting to the realities of war, making arrangements, striking bargains" (46).

34 Even Walzer (2004) offers some of the self-imposed restrictions the air combatants imposed on their pilots during the air campaign of the Gulf War as an example. He noted that not only were discrimination and proportionality discussed in operation planning, but that they were also in practice.

35 See Regan (1996, 99). Anthony Hartle (2002) also discusses the how effective the rules of war have been in spite of incidents like I have mentioned.

36 Christopher (2004) argues, "That political leaders can create circumstances that lead to unjust fighting does not, however, imply that soldiers qua soldiers have no responsibility under the just war tradition. They bear heavy responsibilities, but not of a jus ad bellum nature. The responsibility for the conduct of the war is, as Grotius notes, always a military one. In fact, all soldiers are always morally and legally responsible for their actions on the battlefield, regardless of whether the actions are based on superior orders" (89).

37 For an interesting study into the killing psyche in war, see Grossman (1996), Holmes (1986), and Marshall (1947).

38 Holmes refers to this concept as "just necessity."

39 Also see Hartle (2002) for his discussion of atrocities and dehumanization in war, including during World War I (965–6).

40 DDE is commonly attributed to Aquinas, *Summa Theologica* (II-II, Qu. 64, Art.7). See also Anscombe (1970, 42–53), Philippa Foot (1967), Benbaji, (2008c, in particular 475), Walzer (1977, 153–5), and Christopher (2004, 92–5).

41 For a good discussion of this and other problems of DDE, see Alastair Norcross's "The Road to Hell."

42 See Walzer (1977, 41) and Bomann-Larsen (2004, 142–4).

43 "This apparently represented the highest recorded percentage of contractors used by DOD in any conflict in the history of the United States" (Aftergood).

44 See Regan (1996, 87–95), Holmes (1989, 104), Hurka (2008, 135–6), and Bomann-Larsen (2007, 39–40; 78–89).

45 I do not intend on a full discussion of what qualifies one as being "innocent" here. Part of the problem of identifying a non-tendentious definition of "innocent" is that it seems predicated on the necessary and sufficient conditions that allow one to be morally targeted. So, for example, if "engaging is harm" is the condition for morally permitted targeting, then "innocent" might simply mean: one not engaging in harm. But, this definition is too encompassing. What about the sleeping soldier or the cadet in training? Furthermore, we accept the idea that certain civilians can be targeted; they are not innocent in this sense. Plus, they are materially contributing to the war effort. The legal definition is sufficient for our work here.

46 Geneva Convention IV, "Geneva Convention Relative to the Protection of Civilian Persons in Time of War," Art. 3.

47 Also see Walzer (2006b, 3–4) and Benbaji (2008c) for an in-depth discussion of this issue and why they remain targetable.

48 Bomann-Larsen (2004, 144; 2007, 37) notes that combatants "include all members of the regularly organized armed forces of a party to the conflict (except medical personnel, chaplains, civil defence personnel, and members of the armed forces who have acquired

civil defence status), as well as irregular forces who are under responsible command and subject to internal military discipline, carry their arms openly, and otherwise distinguish themselves clearly from the civilian population." She draws this from http://www.cpf. navy.mil/pages/legal/NWP%201-14/NWPCH.htm, accessed on 23 May 2003.

49 McMahan (2008) writes, "Some combatants—the ones who fight for a just cause within appropriate constraints—retain all their rights and are therefore innocent in the relevant sense, while some non-combatants bear a significant degree of responsibility for a wrong the prevention or correction of which constitutes a just cause for war; they may therefore be liable to certain harms if harming them would make a proportionate contribution to the achievement of the just cause" (27). I think that McMahan is conflating noncombatants and non-innocent civilians. But, in reality, this distinction is very difficult to parse.

50 Holmes (1989) does attempt to draw this line. He writes, "Members of a nation . . . will fall roughly into one or more of six categories: (a) initiators of wrongdoing (government leaders); (b) agents of wrongdoing (military commanders in combat soldiers); (c) contributors to the war effort (munitions workers, military researchers, taxpayers, etc.); (d) those who approve of the war without contributing in any significant way; and (e) non-contributors and non-supporters (e.g., a young children, some active opponents of the war who refuse to pay taxes, the insane, etc.)" (187).

51 What I have in mind here echoes, I think, Anscombe (1981) in saying that farmers are not CIV_{WE}, but those persons "assimilated to the class of soldiers—partly assimilated, I should say, because these are not armed men, ready to fight, and so they can be attacked only in their factory (not in their homes), when they are actually engaged in activities threatening and harmful to their enemies" are CIV_{WE} (67).

52 Also see Scahill (2007, 117–32), Pelton (2006, 154–65), Prince (2014), and Bremer and McConnell (2006, 316–27). There are of course many other anecdotal examples of contractors involved in combat, including Ashcroft and Thurlow (2006, 169–91) and Schumacher (2006, Chapter 7).

53 Department of Defense, Defense Acquisitions Regulations System, Contractor Personnel Authorized to Accompany US Armed Forces (DFARS Case 2005-DO13) Interim Rule, Federal Register, volume 17, number 116, 16 June 2006, 34826-7.

54 The updated version is ADP 3-0: *Unified Land Operations* (2011). It notes that "Offensive operations are operations conducted to defeat and destroy enemy forces and seize terrain, resources, and population centers. They include movement to contact, attack, exploitation, and pursuit. Defensive operations are operations conducted to defeat an enemy attack, gain time, economize forces, and develop conditions favorable for offensive and stability tasks. These operations include mobile defense, area defense, and retrograde" (5–6).

55 Percy (2007b) points out several other loopholes in Article 47. She writes, "First, Article 47 2(b) appears to allow significant 'wiggle room' for states to use private force *themselves* while making sure it remained illegal for other actors. Second, Paragraphs 2(b) and (f) seem to be an intentional loophole in the law that would allow states to continue to use foreign technical advisors, which some critics argue demonstrates a striking lack of enthusiasm for genuine anti-mercenary legislation" (170). She continues, "Article 47 (2)(e) indicates that states are allowed to use private force as long as it is integrated into their own forces, so avoiding the 'mercenary label'" (171). And, interestingly, "Without paragraph (2)(e), venerable military traditions, like the British use of Gurkhas, the Vatican's Swiss guards, and the French and Spanish foreign legions would be considered mercenaries when they previously had not been so considered" (173).

56 See Singer (2003, 97; 208), Silverstein and Burton-Rose (2000, 181; 184; 186), Avant (2005, 233), and Isenberg (2009, 93).

57 From this bid solicitation, the U.S. awarded the $293 million contract to Aegis Defense (Avant 2005, 226–7).

58 Cf., Byers (2005, 119).

59 As Avant (2005) also observes, "Also many jobs not typically considered core military tasks nonetheless have taken on greater danger or ability to inflict harm when performed in the midst of an insurgency. Truck driving may not sound like a core militaries responsibility, but KBR employees have died transporting fuel to troops when they have had to pass through combat zones to get there" (21–2).
60 Ironically, the PSC contractors who are employed by the State Department often enjoy immunity from local prosecution, like in Iraq before 2008.
61 See Singer (2006) and Beah (2007).
62 I would like to thank members of the audience, especially Fernando Teson and James Pattison, at the 2010 International Studies Association (ISA), February 2010, New Orleans, as well as Richard Schoonhoven, for their comments on an earlier version of this chapter.

6 The challenge of military privatization to the military profession

> The modern officer corps is a highly professional body. It has its own expertise, corporateness, and responsibility. The existence of this profession tends to imply, and the practice of the profession tends to engender among its members, a distinctive outlook on international politics, the role of the state, the place of force and violence in human affairs, the nature of man and society, and the relationship of the military profession to the state.
>
> (Huntington 1957, 7)

Militaries around the world today consider themselves professional; that is, they consider that they are members of the profession of the military. In particular, the US military views itself as a "profession of arms," similar to other professions in society. Like other professions, such as law or medicine, the US military believes it serves a specific function or role for society; namely one hears that the public has entrusted the military—and only the military—with the responsibility for the defense of the state. More importantly, it is this very responsibility, "to support and defend the Constitution," that has increasingly, although perhaps unintentionally, been eroded. This erosion is not due to incompetence of the military, nor am I speaking of sequestration and budget reductions. Rather the culprit is the outsourcing to and increased employment of armed contractors. The hiring of armed contractors is not a new phenomenon; however, the increases of both the numbers of armed contractors and the acceptance of expanded roles in foreign policy since 9/11 have not only resulted in an acquiescence to their larger role in military deployments. The continued employment of and reliance upon armed contractors also forms a threat to the military as a profession.

"How," one might ask, "are private military companies challenging the military profession?" Consider the following example. In 2006, then Blackwater vice-chairperson, J. Cofer Black, offered Blackwater's services as a peacekeeping force, not just in an advisory or training role. Rather, Black proposed a brigade sized, contracted force that could intervene, provide security, and ensure the safe distribution of international aid to conflict-ridden areas such as Darfur (Isenberg 2008).[1]

Normally, however, peacekeeping is a role of the military, and these missions would fall under the Operational Support category for PMCs. As of 2015, the UN has conducted (or is conducting) seventy-one peacekeeping operations supported or lead by military forces, and in July 2015, there were 16 ongoing UN peacekeeping missions, with over 107,000 uniformed personnel (UN 2015). As Dag Hammarskjöld famously said, "Peacekeeping is not a job for soldiers, but only soldiers can do it." For the US military, peacekeeping missions fall under Stability and Support Operations, and as such form part of the military's core competencies.[2] Recently, the US military has participated in many such humanitarian interventions (e.g., UNOSOM I and II in Somalia), and it has participated in humanitarian activities domestically, such as the Hurricane Katrina relief operations. (But then, so did Blackwater and other PMCs (Sizemore 2005; Blackwater USA 2005; Witte 2005).)

Nevertheless, for each of these humanitarian or peacekeeping missions, other regions have suffered without intervention, for example Rwanda, Burma, and Darfur. The Syrian conflict today has resulted in a growing humanitarian crisis. Perhaps, as some have wondered, a lack of ability or will to intervene has prevented the UN from taking action; the UN of course has no standing forces of its own. Here then is the profitable potential offer: for a fee, PMCs would intervene on behalf of the UN (or some other presumably legitimate authority) to secure the peace.[3] Max Boot (2006) notes, "Yet this solution is deemed unacceptable by the moral giants who run the United Nations. They claim that it is objectionable to employ—sniff—mercenaries. More objectionable, it seems, then passing empty resolutions, sending ineffectual peacekeeping forces and letting genocide continue." Rebecca Ulam Weiner (2006), Fellow at the Kennedy School of Government, seems to reinforce this view. She writes,

> The discomfort [of using PMCs for peacekeeping] also has deeper roots, in the complicated history of private intervention in these kinds of conflicts. When Kofi Annan was UN undersecretary general for peacekeeping, he explored the option of hiring the South African private military company Executive Outcomes to aid in the Rwandan refugee crisis. He ultimately decided against the option, declaring that "the world is not yet ready to privatize peace.

Michael Walzer also entered the discussion. In his *New Republic* article, "Mercenary Impulse," Walzer (2008) considered whether it might be useful to employ what he terms mercenary forces for peacekeeping missions where the UN and the rest of the international community are unable or unwilling to intervene. Even actress Mia Farrow petitioned Blackwater in 2008 to help in Darfur (Morris 2008).

Yet, the hiring of these private forces, even for peacekeeping operations, sounds magnanimous; it is not so different in kind from the hiring of private armies that international anti-mercenary laws seek to prevent. Furthermore, from the US and coalition forces in Iraq battling sectarian, private militias (e.g., Mahdi Army, etc.), to the fight against ISIS today, to other governments' forces combating organized crime and drug cartel private militaries, states (1) recognize that these private militaries are destabilizing to a region, and (2) are seeking their elimination or assimilation. This policy contradiction highlights one way in which the employment of armed contractors challenges the military as a profession. To explore this issue in greater detail, I will first analyze the concepts of professions as well as the military as a profession. I draw upon Samuel Huntington's (1957) and others' explorations of the attributes of a professional military,[4] and I argue that, according to these conditions, the military is a profession. Next, I contrast these conditions of the professional military with those of privatized military. I argue that, although there are many similarities, and perhaps some conditions overlap, the professional military is distinct from PSCs and PMCs. Armed contractors, while perhaps professional in some sense, are not members of a profession. I then draw upon these distinctions to demonstrate how PSCs and PMCs broadly are incompatible with professional militaries and how they pose a serious challenge to the military as a profession. One might counter, "Who cares if using PMCs erodes the military of a profession?" "Perhaps," she might add, "the use of PMCs might be more efficient or cost effective." The problem is not, however, limited to the military itself. An erosion of the society's military profession affects the civil-military relationships of that society. How to fight wars and the decision itself to go to war itself would change. More specifically, I offer that this erosion on the identity and authority of the military profession negatively affects public policy, including the jus ad bellum decision to go to war.

PMCs are not members of the military profession

Recall from Chapter 3 the following explication of "Professional Soldier:"

Professional Soldier: S is a professional soldier if and only if

1 S's primary occupation is soldiery, a specialized expertise attained by prolonged education and experience;
2 and S
   a is a citizen or resident of the state in which she is fighting;
   b is integrated into a national force; and
   c is primarily motivated by her civic duty to the state (even if she also is financially compensated); and
   d is a member of a military profession, whereby the profession exhibits:
      i a specialized expertise in warfighting;
      ii a responsibility to provide national security; and
      iii a sense of professional corporateness; and

   e swears an oath as a government appointee, embracing the duties and responsibilities as a member of the military profession.

The definition of a PSC contractor is as follows, which is derived from the definition of a PSC from Chapter 3:

PSC contractor: S is a PSC contractor if and only if

1 S is employed by either a publicly-held or private business company (a PSC) that:
   a is organized along corporate lines;
   b is not integrated into a national force;

   c   operates from a business profit motivation;

   d   is a legal entity operating on the open market as its own entity or through a parent corporation;

   e   is organized into an established, permanent entity; and

   f   offers a range of military related, Policing/Security services (e.g., personal security, site or installation security, convoy security, and policing); and

2   S is hired by the PSC to provide security and combat related skills.

The above definitions capture the two areas, in particular, that on the one hand are sometimes shared between the US professional soldier and PSC contractor, yet on the other hand, also draw out certain, important distinctions. First, the professional soldier swears an oath. While perhaps similar in form to the volunteer codes of conduct of the private security industry (e.g., the ISOA Code),[5] the professional soldier's oath alone captures the underlying obligations entailed by becoming a member of the military profession, and the legal and professional responsibilities of government agents. Verkuil (2007, 112) notes, "The [military] oath is not a mere formality. It separates public and private actors. It is something that government officials have in common …."[6] The oath also serves as a continuance of the executive authority as an agent of the government:

> If the president assigns duties to private contractors that are normally performed by either principles or "inferior" officers of the United States, the vertical dimension of separation of powers is triggered. Officers of the United States "exercis[e] significant authority"; this is authority inherent in executive function. Transfer of this function to private hands potentially violates authority delegated to the executive.
>
>   (103)[7]

Secondly, while both the professional soldier and the PSC contractor sign contracts—that is, they are contractually obligated—the PSC contract captures only the business and perhaps legal obligations of the employee and her employer. As I discussed in Chapter 3, the soldier also signs a legal contract with the state, but she further enters into a different kind of contract—an implicit contract between the military and the society it serves. Martin Cook (2004) argues that in the US military,

> The terms of the contract are that the military officer agrees to serve the government and people of the United States. He or she accepts the reality that military service may, under some circumstances, entail risk or loss of life in that service. This contract is justified in the mind of the officer because of the moral commitment to the welfare of the United States and its citizens.
>
>   (74)

Furthermore, these definitions draw comparisons to individual roles. A PSC or PMC is a collective business entity. So, while the contractor herself might be motivated by her patriotism and may be willing to accept severe risk like the soldier, the company and the contract are independent of her beliefs and role in a way that the professional soldier's is not.

While many would simply agree that PSC contractors are different from professional soldiers, the differences rely upon a key assumption: that the military is a profession. Therefore, in order to not only demonstrate the distinctions between PSC contractors and professional soldiers, but to discuss the effects of military contracting on the professional military, it would be useful to illuminate the characterization of "profession" being assumed.

We have all heard various descriptions of employment. Is it a job? A vocation? A tradecraft? A calling? A profession? While each of these descriptions implies using labor to provide goods or services, for our purposes, we need a workable conception of "profession." There are of course differing ways to explicate a profession, but the majority of them attempt to identify the attributes or characteristics that differentiate the profession from a typical "nine to five" job. Samuel Huntington famously proposed, in his 1957 *The Soldier and the State*, that professions are characterized by their expertise, responsibility, and corporateness (8). However, there have been many different attempts to identify the distinguishing characteristics of a profession. For example, consider a sociological approach. Ernest Greenwood in *Man, Work, and Society*, writes that according to this view, a profession is "constantly interacting with a society that forms its matrix, which performs its social functions through a network of formal and informal relationships, and which creates its own subculture requiring adjustments to it as a prerequisite for career success" (quoted in Hartle 2004, 21). While perhaps interesting, the sociological account does not seem as applicable to the debate at hand; PSCs also function through "networks of formal and informal relationships."

Alternatively, James Burke (2002) recognizes three factors that make an occupation a profession: expertise, jurisdiction, and legitimacy. Philosopher and retired Brigadier General Tony Hartle (2004) suggests that the attributes of a profession include (1) systematic theory, (2) authority, (3) community sanction, (4) ethical codes, and (5) distinct culture (22). Military historian and theorist Allan Millett (1977) further offers a more complete account, which accounts for these differences. An occupation that fulfills these conditions is a "Millett's Profession":

Millett's Profession: an occupation O is a profession if and only if O

1   is a full-time and stable job, serving continuing social needs;
2   is considered a life-long calling by the practitioners, who identify themselves personally with their job subculture;
3   is organized to control performance standards and recruitment;
4   requires formal, theoretical education;
5   has a service orientation in which loyalty to standards of competence and loyalty to clients' needs are paramount; and
6   is granted a great deal of collective autonomy by the society it serves, presumably because the practitioners have proven their high ethical standards and trustworthiness (derived from 2).

Millett's Profession seems to capture the essential elements of what a profession is and how it differs from the employment of PSC contractors; however, it is somewhat unwieldy for this task and may lengthen our professional soldier definition unnecessarily. Martin Cook (2004) offers a shorter list of attributes, which he notes as professional knowledge, professional cohesion, and professional motivation and identity (66). Note the commonalities.

Nevertheless, I think that Huntington's conception remains both the simplest and relevant for this discussion. As I noted in the definition of a professional soldier, Huntington's characteristics of expertise, responsibility, and corporateness are captured by "(i) a specialized expertise in warfighting; (ii) a responsibility to provide national security; and (iii) a sense of professional corporateness." Furthermore, the professional soldier's characteristics also capture the concerns highlighted in the Executive Summary from the US Naval Academy's 9th Annual McCain Conference on Ethics and Military Leadership ("Ethics and Military Contractors" 2009). In referring to the use of PSCs in both conventional and irregular war, it reads,

> The use of deadly force must be entrusted only to those whose training, character and accountability are most worthy of the nation's trust: the military. The military profession carefully cultivates an ethic of "selfless service," and develops the virtues that can best withstand combat pressures and thus achieve the nation's objectives in an honorable way. By contrast, most corporate ethical standards and available regulatory schemes are ill-suited for this environment.[8]

Part of what constitutes a profession's expertise, in our case the specialized expertise in warfighting, is as Alan Goldman (1980) notes in *The Moral Foundations of Professional Ethics*, "the application of a specialized body of knowledge in the service of important interest of a clientele" (18; quoted in Hartle 2004, 10), (or according to Millett (1977)—a formal, theoretical education). In the military case, the knowledge refers to the management of violence, and, in the United States, American society is the client. One of the areas where the effects of hiring privatized military are manifested is in the acquirement, development, and education of this specialized body of knowledge.

How the employment of PMCs affects the military profession

> A king who relied too much on mercenaries risked not only treachery but also was liable to alienate his barons, as King John did using men such as Fawkes de Breauté.
>
> (Barber 1980, 14)

To my knowledge, there are no official, systematic studies as to the effects of hiring private military on the military as a profession; although there are a growing number of attempts to explore parts of this issue (Heinecken 2009; 2014; Schaub and Franke 2009). But, that does not mean that the effects are negligible. Some of these effects may be psychological, affecting how the professional military as an institution views itself and its place in society—its identity. For example, whole scale privatization of the military may eliminate the military's professional nature, but currently complete privatization seems unlikely. The question is how may an incremental privatization affect the military as a profession. On an extreme view, one posited by Singer (2003), is that if the military's corporate identity is threatened enough, it could "be an impetus for action against its own regime" (198). While I think that the dire consequence of a military coup on these grounds is implausible (at least in the United States and other Western states),[9] it has happened elsewhere. Short of a coup, since certainly the corporateness of the military profession is an essential element, its corporateness will undoubtedly be altered. Consider how the reaction voiced by the Papua New Guinea military against its government's hiring of Sandline reflects this psyche (Cockayne 2007, 211; Singer 2003, 197; Percy 2007b, 211–12; Axelrod 2014, 217–20).[10]

Furthermore, in the United States, the military consistently remains one of the most (if not the most) respected governmental institutions in the eyes of the American people, at least according to recent Harris Polls (Volcker 2003). Huntington (1957) writes that the professional soldier's "behavior in relation to society is guarded by an awareness that his skill can only be utilized for purposes approved by society through its political agent, the state" (15). When the purpose of the institution or its authority no longer remains or is lessoned, the profession will transform or be eliminated, and civil-military relations will change. This kind of effect is of course more difficult to quantify; nevertheless, these effects may be felt in other shifts to military privatization—ones that are more apparent. In this section I will look at how military privatization affects a loss of capacity and experience, the friction caused by adding financial incentive to the profession, as well as how the guardians of the professional military's specialized body of knowledge has changed. As Avant (2005) points out, the professional military seems to be "ceding control over the shape of a core element of its professionalism to commercial firms that may be influenced by a variety of consumer demands rather than the US government's alone" (120),[11] or, perhaps more relevantly, the people's demands.

Effects on the professional military's expertise

Loss of capacity and a loss of experience

In the logistics world of the military, cooks, truck drivers, and mechanics have been outsourced to one degree or another (e.g., part of LOGCAP, discussed earlier). As many have admitted, the military cannot deploy without the PMCs. Consequently, much of the military's logistics capacity has shifted to private hands. This privatization has also had a different, perhaps unintended effect. By shifting certain capacities to outside of the military, there is a corresponding loss of internal experience and expertise in the planning and employing of logistics assets. (Of course, the logistics commands would argue otherwise, but this view would be shortsighted, as the current senior logistics officers served as junior officers in a military that previously had these logistic capacities internally.) In the future, there may be an expertise gap, which could then be filled only through contracting of yet another military service—the management of logistics. This move raises two other issues: first, PMCs in the logistics arena will have to look elsewhere, outside of the military, for their employees; and second, the military may eventually have to outsource all logistics, to include the policing of logistic functions due to a lack of experience.

Whether intentional or not, the military's outsourcing of logistics entails certain feedback effects upon the military. One goal of privatizing cooks, for example, is to increase the number of combat forces, which is arguably more important than military culinary skills. Nevertheless, the feedback effect is that cooking itself is no longer important to the military; therefore, the military simply focuses less on cooking. The expertise now lies outside of the military institution. (A similar out-sourcing transformation occurred with the Army's Stryker vehicle (GAO 2006).)

Cooking skills are one matter; take a different example: the training of foreign military forces. As I noted earlier, the effort that a state expends on training foreign military forces can pay large dividends, serving its international interests. Never-theless, states, such as the United States, are increasingly turning to using private military for these training missions. It is wise to consider these implications as we look to expand both the advisory role in Iraq and the training of moderate Syrian opposition forces. There are certainly positive trade-offs when hiring former mil-itary personnel to conduct the training missions, but the feedback effect, in terms of how the military itself now views the training mission, has had a negative effect.

The hiring of DynCorp in Bosnia and Vinnell in Saudi Arabia has the unintended effect of reinforcing the message to soldiers that training is not as important—not a core mission. Why should military members volunteer or seek out these kinds of training missions? In the US Army for example, it was a common (perhaps unfair) understanding among junior leaders that training commanders were viewed as somewhat less capable of future potential than "line commanders." While not the official policy, officers tended to avoid training assignments. In Iraq and in Afghanistan, however, there was a renewed emphasis in the training of other forces and a corresponding interest in developing appropriate military skills and attracting proficient leaders to undertake these now vital training missions. Directions to the Army's promotion boards, for example, were emphasized to give those selected for training assignments an equal opportunity for advancement—to change the negative stigma of training missions. Training the Afghan military requires skills that are in short supply in the military; even internally in the US Army, units like ours that rotated into Iraq underwent "certification drills" and training in Kuwait, conducted by MPRI and other PMCs. It was as if the Army thought that additional layers of training and certification were essential to ensure that all rotating units met minimum theater-specific proficiencies, and it turns to private companies to help.[12]

In summary, the outsourcing of a certain skill causes feedback effects that (1) the skill is no longer viewed as vital within the military; (2) expertise of that skill is now developed outside of the military; (3) the pool of former military personnel with this skill will shrink; (4) should the military require the skill once more, it will then have to turn to the private sector; and (5), as Avant (2005) cautions, "Outsourcing training also opens the way for changes in the hold of public institutions on the development of military norms, changes in the values attached to military service, and changes in policy resulting from a different policy process" (142).

This feedback effect loop becomes more troublesome in the security arena. Ironically, "[PSCs] market the unique expertise that their employees gain from service in the publicly funded military" (Singer 2003, 205). PSCs need former military personnel and expertise, and in places such as Iraq and Afghanistan, the military increasingly seems to need PSCs to do some of the security tasks that would normally fall to the military itself. One might argue that PSCs are not needed because of the changing nature of war; rather, our own desire for efficiency and supposed cost savings has made PSCs necessary. The longer that a state relies upon private military services, the less the state will be able to conduct the services on its own. However, without the pool of experienced military personnel that PSCs rely upon, the security companies may not be able to provide the very services that caused the states to turn to them in the first place.

The military profession's education

There is another area where the privatization of military functions has had an effect on the profession's expertise—education. Since 1997, the US Army has staffed much of the teaching responsibilities of its Reserve Officers' Training Corps (ROTC) programs to private contractors. While it seems an innocuous move, the privatization of ROTC has generated some controversy. Some of the controversy, of course, surrounds whether the use of private contractors in ROTC programs actually saved costs as intended (Verkuil 2007, 177; Avant 2005, 116–20);[13] indications are that it does not.

However, this program generates a deeper concern: namely, the potential shift of the responsibility for training the future leaders of the military profession outside of the profession itself. Referring to how an individual develops expertise as a member of a profession, Huntington (1957) writes

> [A professional's] education consists of two phases: the first imparting a broad, liberal, cultural background, and the second imparting the specialized skills and knowledge of the profession . . . . The second or technical phase of professional education, on the other hand, is given in special institutions operated by or affiliated with the profession itself.
>
> (8–9)

It is the second phase of professional, technical education that is worrisome. Hartle (2004) notes that, "One gains professional expertise through a lengthy period of formal education." Furthermore, he adds that "Usually, the profession itself controls and perpetuates professional education to its own facilities" (11). Additionally, as Cook (2004) observes, professional expertise is further developed through "extensive periodic education and training designed to impart professional knowledge and expertise, and they practice the application of that knowledge at all levels of their work—from field training exercises, through National Training Center rotations, to strategic exercises and wargames" (63).[14]

For the Army officer, especially, this process begins as she starts her professional education through one of the commissioning sources (Officer Candidate School (OCS), West Point, or through ROTC). It is this initial exposure to the profession that is critical to her future development. My own initial exposure to the military profession began 30 years ago as a New Cadet at West Point. Not only did I learn the basic technical skills of the Army, I also learned of the Army's "unique character of military service and the peculiar sphere of military competence: the systematic application of force for political purposes" (Hartle 2004, 13). It was also my introduction to the traditions, customs, and history of the Army. It is where I learned what military service means, how civil-military relations work, and the fundamentals of the oath of office I eventually took as an Army officer and what responsibilities I would inherit. At West Point, I also learned about the laws of armed conflict and bellum justum, two areas I continue to practice and teach today. As Avant (2005) points out,

> Western military personnel see it as obvious that they should abide by the obligations of international law with respect to the conduct of war and domestic law with respect to civilian authorities. These requirements are embedded in professional military education not only as moral obligations but also as integral to effective military operations.
>
> (52)

Many of these same skills and lessons can be taught by others: former military personnel and civilian faculty, with little loss in effectiveness. Supporters of privatizing ROTC counter that the companies hired only use former military personnel, and they work close in hand with the military in developing the curriculum and maintaining current with evolving military matters. Further, the military itself still maintains control, oversight, and overall responsibility for ROTC.

While these observations may be true, by outsourcing even parts of an officer's education, the Army is ceding some control over its own expertise to outside sources —ones that, despite close military ties, are regulated by the market. Another concern is how to correlate the education of professional military ideals such as selfless service when these concepts are being taught by for-profit educators. Burke (2002) notes that this program of privatization "may suddenly teach new entrants into the profession that, despite the rhetoric of self-sacrificing leadership, market logic trumps other considerations" (35). Some have hinted the worry that a market-based approach to military education might erode the very republican ideals of service to the state to the notion of sacrifice for the greater good, the society, for others—all ideals the professional military attempts to inculcate and foster. (Although, I think this is an overly pessimistic view.)

Even outside of ROTC, the privatization of military education is evident. Huntington (1957) observed that "Professional knowledge, however, is intellectual in nature and capable of preservation in writing" (8). Yet, much of the military profession's own writing is subject to outsourcing. A PMC, Cubic, was hired to transform the curriculum of the Command and General Staff College (CGSC) into the Intermediate-Level Education (ILE) program in 2001–2003 (Cubic" 2003).[15] Moreover, private contractors are used to assist in the writing of military doctrine, including the doctrine that covers the employment of private contractors. (See FM 3-100.21 *Contractors on the Battlefield* (2003c) and FM 4-100.2, *Contracting Support on the Battlefield* (1999).)

Even proponents of PMCs, such as Carafano (2008), warn about the professional military's loosening control over its own education, especially when the education concerns preparing for the next war. He writes, "The Military has a mission. It is the military's job to think about the future of war .... Among the tasks the military cannot contract out is responsibility for facing the future of war" (205).[16] While the effects of PMCs on the military profession's expertise may be incremental, there is another area where the profession feels the effects. There is an increasing worry that the military's corporateness is also being challenged.

Effects on the professional military's corporateness

> Men can be motivated by money to undertake dangerous and irksome tasks, but the result would be to weaken essential heroic traditions. In a private enterprise society, the military establishment could not hold its most creative talents without the binding force of service traditions, professional identifications, and honor.
>
> (Janowitz 1960, 422)

Professional military corporateness is defined by the certain principles and ideals of the professional military, derived from its role of service to its society. Corporateness is also defined by the identity of membership of those who belong to the military profession. Christopher (1999) notes that members of the military profession have "a deep commitment to a set of abstract values and principles that define the profession." Furthermore, "members of a profession accept [these] values that are specific to their profession as being more fundamental than other values. [e.g., the Hippocratic oath]" (212–13). History and tradition, as Janowitz identifies, play large roles in developing within the profession an attachment to membership within the profession and a "commitment to the role entails a 'profession' of obligation to society" (Hartle 2004, 12).

There are two areas in particular were increased hiring of PMCs affects the military corporateness. First, the practice of using market-based employees challenges the profession's ideal of its social obligations to the state. Second, the use of PMCs, although perhaps less measurable, affects the mindset of the military members of the profession. In other words, how a soldier grounds her professional identity may be changing. Part of the military profession's corporateness, tied to its responsibility to the society it protects, is that the day-to-day responsibilities constitute more than a mere job. Rather, the professional military sees it as a calling or vocation not tainted by commercialization. Charles Moskos, in his "From Institution to Occupation," worried "that a possible turn toward viewing the military function as a job will negatively affect the effective carrying out of the military mission" (Wakin 1979, 7). This integral conception of the military profession as a calling, not guided by profit incentive, is partially rooted in theory. As Huntington (1957) notes, "Financial remuneration cannot be the primary aim of the professional men qua professional man. Consequently, professional compensation normally is only partly determined by bargaining on the open market and is regulated by professional custom and law" (9). It is also tied to pragmatic concerns as well. As I discussed earlier, professional soldiers do enjoy some financial compensation. However, as Janowitz (1960) argues, "In the long run, it is doubtful whether the military estab- lishment, like other public agencies, could maintain its organizational effectiveness merely by raising monetary rewards, and by making the conditions of employment approach those found in civilian enterprise" (422).[17] To be fair, a private force can be effective, especially in the short term. Furthermore, as we explored earlier, the question of effectiveness may be tainted by the historical bias against the hire of private forces in general.

What is clear, nevertheless, is that PMCs as corporate entities are financially motivated, and this profit incentive[18] could conflict with the needs of the society. Furthermore, there is an intuition at work as well. The professional military serves to provide security to a society just in case the society is threatened. PMCs, alter- natively, seem to require conflict (or the threat of conflict, much like Eisenhower's military-industrial complex did; I will turn to this later) to generate profits. Without conflict—without the need for increased security—they will have to transform into a different industry.

Advocates of PMCs would counter, as Carafano (2008) argues:

> It is wrong to assume, as critics so often do, that [contractors] are not bound by codes of conduct as well, codes that keep them from putting patriotism over profit. They are sanctioned by the kinds of codes that Adam Smith first envisioned as the best strictures of moral behavior for governing a free market—codes that are informed by and spring from enlightened self-interest.
>
> (165)

However, enlightened self-interest might (1) demand setting aside the voluntary PMC industry code and choose profit over patriotism, and (2) the globalization of the PMC industry will result in contracts where differing interests diverge. As Leander (2007) further notes, "it is not clear that PMCs' organizational culture is converging with the professional culture of public armed forces, promoted by states ... " (62).[19]

Not only does the profession's conception of its services as a calling clash with the for-profit incentive but the soldier's professional identity itself is also affected. The US military, for example, is often conflicted over recruitment strategies (e.g., the military services' different recruiting commercials)—moving from highlighting the profession's calling and duty to serve the country to how the military can provide ideal training and skills needed for future employment in the corporate world.

Yet, as Wakin (1979) points out,

> When one feels his daily work constitutes a vocation, he gives to it not merely the majority of his time; he in a real sense becomes identified with his work; his whole personality development is associated with what he does and how he does it.
>
> (7)

Cook (2004) adds:

> There is no substitute for the fundamental mindset that members of the profession, regardless of rank, are colleagues, engaged in a common enterprise that matters deeply to them. If that mindset is present, then each member feels a loyalty to the other, grounded in his or her common professional identity. If each thinks of professional identity in this way, each takes pride in responsibility in preserving, developing, and transmitting the body of knowledge that resides at the core of the profession.
>
> (72–3)

Being taught by contractors, trained by contractors, then working with contractors— often former military, many known by their former colleagues—and seeing other contractors not sanctioned for improper conduct, all begin to chip away at the soldier's professional identity and the profession's corporateness. If the ideals of the profession begin to be called into question or are replaced by a for-profit mindset, membership in the military profession may lose its special status, and the relationship between the military and its society will change.

Benefits of PMCs to the military profession

There are certainly benefits associated with use of PMCs. Potential cost savings come to mind, but like the ROTC programs, the use of hired, former military personnel may also allow the military to free up the officers and NCOs, who would normally fill the roles, to work elsewhere, such as in Iraq or Afghanistan. In this section, we will look at the effects of military privatization on emerging professional militaries as well as the benefits of keeping expertise in the military system.

What about von Steuben?

The Revolutionary American Army employed officers from European armies to train the colonies' developing forces. Without their assistance and, more important, their expertise, some question whether the American colonies could have won their independence against the more experienced British Army. Likewise, the use of private military to train other forces not only results in a manpower savings (e.g., the Army has 45 active brigades commanded by colonels—how many units would have to stand down to field a training force of colonels to train another country's senior military leadership?)[20] but it is also a way to capitalize on years of experience and expertise. Additionally, these former military personnel bring the history, traditions, and sense of professional service with them, and could impart these to other countries' forces. In this manner, one might argue, PMCs could help a military in an emerging democracy become a professional organization.

The question becomes: Do former military members maintain the same professional qualifications (apart from the technical, military skills) when they leave the military? I think the answer could be "yes," and most often this is probably the case. Huntington (1957) writes,

> The members of a profession share a sense of organic unity and consciousness of themselves as a group apart from laymen. This collective sense has its origins in the lengthy discipline and training necessary for professional competence, the common bond of work, and the sharing of unique social responsibility. The sense of unity manifests itself in a professional organization which formalizes and applies the standards of professional competence and establishes and enforces the standards of professional responsibility.
>
> (10)

PMC proponents would argue that these same standards follow the former military member, even when she accepts employment as a contractor, and many of these companies advertise this same standard-based professionalism to differentiate themselves from other PMCs. The former military contractor herself may even feel vestiges of her former professional military identity and a close bond with those still in uniform.

However, even if she maintains this identity with the military, she is no longer a member of that military—answerable to the profession's sanctions, ethics, and standards of performance (Janowitz 1960, 6). Furthermore, the company that she works for may also demand that she comports herself to those same standards—reinforced by the conditions of her employment.[21] She may hold herself accountable in a similar manner, perhaps in a "professional" manner, but she has moved outside the profession itself. (The current debate about the role of retired generals draws upon this idea.)[22] Consider, as Cook (2004) observes, the distinction between the professional and what he terms the "para-professional." He writes,

> It is also important to note that, in most professions, there is a core of 'real' members of the profession who have fully imbibed this knowledge. But there is also a large penumbra of para-professionals, who are knowledgeable in some areas essential to the profession's jurisdiction, but only some. To use medicine as example once again, phlebotomists and X-ray technicians are essential to the core functions of medical care, but clearly are not members of the medical profession in the fullest sense.
>
> (67)

Former military contractors are more closely identified as para-professionals than as full members of the military profession. Further, there are many roles which they can fulfill and contribute their service to the state, should they be so motivated. (I think they should be.) Therefore, the military needs to find a way to capitalize on this expertise without losing its own professional identity.

Keep experience in the military system

Unfortunately for those who long for a lasting peace, demand for military experience and skills has not abated. This demand has created traction for the argument that hiring of PMCs keeps the military experience "in the system." There are many former military contractors who no longer serve or cannot serve, yet who have developed hard-earned experience and desire to put that experience to work. I personally know many of these former military who are applying their skills in the private military sector. PMCs offer an avenue for employment doing much of the same work these former military personnel enjoyed. Whether it is through the hiring of retired generals as mentors, whose experience often spans 30 years or more,[23] to the ambassador's security detail guard, who cut his teeth as a Navy SEAL, outsourcing allows them to continue to use their skills (and often be paid handsomely as well).

There are several issues with this business model, however. First, questions of the appropriateness of the salaries garnered arise. Second, military skills are traded on the global market, and there are often former members of the military working for the highest bidder—even for more nefarious employers. Third, while PMCs and PSCs offer a way of keeping former members of the military in the system, this process also generates an attraction of talent away from the profession, generating competition for the military's best and those seeking greener pastures.[24]

Finally, the competition over manpower may hide a potential conflict of interest. The arms industry has had a long history of relying upon retired generals to provide their expertise and open doors to lucrative government contracts. Former military personnel have had to tread a fine line, balancing the ethics of serving the interests of the former profession with their new employers'. Now, many companies in the arms industry have expanded to include security services (e.g., L3 owned MPRI, and L3 spun off its government services through a new corporate entity—Engility Corporation), and the same issues are appearing. Former Army Chief of Staff Peter Schumacher said of his concerns over employing private security on the battlefield, "I can see where, on the battlefield, there would be issues that could be problematic in terms of the rules of engagement, what kind of controls were placed on people that are roaming the battlefield." Yet, as Isenberg (2009) notes, "General Shoemaker eventually resolved his concerns, since a bit over two years later he joined DynCorp International's board" (60).

Certainly, the hiring of former military keeps their skills and expertise in the military system and does provide value-added in a variety of areas. However, it would be prudent to also keep in mind the costs of this practice, especially when the security contracting extends beyond the borders (and interest) of the host state. Furthermore, while the goal of future employment (and perhaps even much greater monetary compensation) might motivate some soldiers to continue to serve or to use their expertise as contractors, it also affects the soldier's professional identity with military profession.

PMCs and the jus ad bellum debate

> [T]he commitment (especially the voluntary commitment) of the soldier to
> selfless service of the society and dutiful obedience to constitutionally valid
> authority is the root of the nobility of the profession, and the source of American
> society's trust in and respect for its profession of arms.
>
> (Cook 2004, 61)

In the last section, I discussed the effects of hiring military contractors on the
military as a profession. However, many of these effects could be re-categorized
as jurisdictional concerns—redrawing of lines of responsibility, for example.
These are areas that currently serving military often point to when questioning the
value of hiring PMCs. Of course, these same concerns must also be considered
when governments contemplate outsourcing any military functions. There is,
however, another area where the widespread hiring of PMCs could have a large
effect. The use of PMCs not only affects the profession itself but also affects civil-
military relations; the practice may also affect the jus ad bellum decision to go
to war. Even if one maintains that Just War Theory is ineffective and an artifact
that needs to be updated, she should agree that public participation in a democracy's decision to wage war is paramount. Furthermore, it is this very debate that
shifts, as a new player—the privatized military service industry—joins it.
I explore this phenomenon in this section, specifically addressing the professional
military's advisory role, the often cited concern of a foreign policy by proxy,
and how the private military industry participates in and affects the jus ad
bellum debate.

The military profession's advisory role

One of the traditional responsibilities of the military profession is to provide military
advice to the civilian authority (i.e. the government) (Huntington 1957, 72; 116;
Cook 2004, 79–93).[25] Part of this advisory role stems from the pragmatic estimation
of whether the military force employed and the military option chosen can lead to the
desired political ends and at what cost. Janowitz (1960) notes, "Military leaders must
be prepared to assist in accurately estimating the consequences of the threat or use of
force against the potentials for persuasion and conflict resolution" (417). The
military's advisory role is also important because it reinforces the importance of
civilian control.

As I discussed in Chapter 5, jus ad bellum decisions are political ones, and the
political leadership has the responsibility of deciding when to resort to war.
The military, alternatively, is responsible for the conduct of war (jus in bello). But,
the military plays an important role by informing the political leadership based upon
the military's expertise through its advisory role. Huntington (1957) writes,

> All members of society have an interest in its security; the state has a direct
> concern for the achievement of this along with other social values; but the

officer corps alone is responsible for the military security to the exclusion of all other ends.

$$(15)^{26}$$

He is only partially correct. The military provides the force and expertise and advice. They may provide the means for military security (and the conduct of such); however, in the realm of jus ad bellum, they provide advice.

Of course, the political leadership is free to accept or ignore the military's advice. In the United States, for example, the president can seek advice from anyone he or she chooses. For the military leader, the following dilemma may then arise:

> Consider a circumstance in which professional military advice has been rendered but not accepted by the professional's civilian superiors. Here [one] must decide whether this is the occasion for obedient service in deference to civilian leadership (the normal case) or whether the course of action chosen by political leadership is so at variance with sound professional judgment that conscientious resignation should be entertained as a possibility.
>
> (Cook 2004, 63)[27]

Interestingly, security or international policy consulting and research firms (i.e., think tanks) are increasingly being consulted when making foreign policy decisions.[28] While perhaps not PMCs per se, the think tanks share similar characteristics, including the hiring of former military personnel. Although it certainly appears beneficial to seek a wide range of intelligent advice and information when making high-impact policy decisions, it would also be interesting to research how the practice of using think tanks affects the military's advisory role, especially, when their views diverge.

Nevertheless, society relies upon the military profession in its advisory role. The military is then an important participant in the debate over jus ad bellum decisions. They are not actively or publicly vocal (perhaps there are times they should be); however, based on their expertise, the advice provided by the military leadership generally carries much weight. (Consider GEN Petraeus' much anticipated 2007 report to Congress on the status of the surge in Iraq.) If the hiring of PMCs affects the military profession itself, then will the military's advisory role alter as well?

Foreign policy by proxy? Really?

Several authors, including Avant (2005) and Singer (2003), warned that the hiring of PMCs opens the door for governments to conduct foreign policy outside of the normal checks and balances and governmental oversight—foreign policy by proxy, if you will.[29] While certainly these kinds of secretive measures are possible, especially in weaker states where government control over forces is already limited, it seems less of a concern in stronger, democratic states such as the United States or Great Britain.[30] (Of course, as I have already discussed, there are historical examples of this kind of government circumvention, e.g. such as the Iran-Contra affair, and military operations in Laos and Columbia.)

In weaker states, such as states that lack a professional military, private force may be the only security option. Other states may use a combination of private and public forces. However, as Huntington (1957) argues, the military has responsibility for security of the state, and when the social responsibility is shared between the military and PMCs, the lines of responsibility blur. This blurring of the security role does allow for governmental loopholes.

A simple example is the concept of "troop caps." For a given conflict, the US Congress for example, may limit the number of soldiers deployed to a given area of operations. Yet, the number of PMC contractors allowed is rarely ever discussed. Consider President Obama's 2009 decision to send 30,000 additional troops to Afghanistan. What he did not discuss was the thousands of new contractors (at least 56,000 by one source) that would be required to help support the increase in troops as well as provide security for the increase in other government agencies' personnel (Pincus 2009b). Unlike the numbers of soldiers committed, these contractor numbers are simply outside the debate.

In terms of the functions that PMCs provide, Verkuil (2007) argues that as long as the authority remains with the government officials, the "decision is in a formal sense still governmental." Problems arise when there is a lack of sufficient government oversight. Verkuil continues, "The question becomes: should the exercise of government authority by officers of the United States involve more than rubber-stamping the work of private contractors?" (110). Consider the case of questioning prisoners (or captured terrorists). The lack of a robust interrogation program in both military and CIA led both to contract with PMCs for assistance ("CIA Lacked" 2004). The methods that were employed included harsh interrogation techniques, which raised a lot of moral and legal rancor. Setting aside the question of whether these techniques constitute torture, one can be certain, as Verkuil (2007) aptly writes, "Whatever else it is, torture is surely a governmental 'function' that cannot be privatized" (130).

I agree with Carafano that the US government may not be able to conduct wholescale foreign policy by proxy through PMCs. However, the concern remains that as lines of responsibility between what is the military's purview and what is the responsibility of PMCs blurs, the military profession's responsibility to society may change or may become less clearly defined. Governmental transparency seems to be a fundamental tenet in working democracies. A blurring of security responsibility, even unintentionally, affects this provision.

PMCs join the jus ad bellum debate

The interest of the profession requires it to bar its members from capitalizing upon professional competence in areas where the confidence has no relevance and likewise to protect itself against outsiders who would claim professional competence because of achievements or attributes in other fields.

(Huntington 1957, 10)

Consider democratic states that hire PMCs from within their own country. The PMC employees here have always been a part of their state's jus ad bellum debate, assuming they have been active participants in the political process. Professional soldiers from these states also have the opportunity to be politically active through voting. In this sense, every eligible voter in a democracy can be part of the ad bellum debate, even if in only a small way. Certainly, one citizen's vote has very limited effect, even when it is aggregated with others' votes. Democracies, too, have conducted foreign policy without consent or over strong protest. Ideally, however, the citizen's voice counts in a democratic society. Thus, the soldier from such a society maintains her autonomy as a citizen as well.

As was discussed, the professional military also serves its advisory role in the decision to wage war, a role it traditionally has had. As the PMC industry grows, it also participates, but now as a corporate identity. Certainly, PMCs have some influence in current security decision making. Some claim, for example, that former Vice President Cheney favored KBR as a logistics PMC because of his former Halliburton ties; this may or may not be true. What is unclear is whether KBR influenced Cheney's policy views. The question at hand is whether the private military industry may displace the military in its advisory role, and if it does not, then in what way is the ad bellum debate affected?

Singer (2003) observes that PMC employees, "are directly responsible to the corporation and its executives; they are hired, fired, promoted, demoted, rewarded, and disciplined by the management of the private company, not by government officials or the public" (154). With the military, there are differing "motivations, responsibilities and loyalties" at play. According to Hartle (2004), because the professional military is traditionally subordinate to civil authority, "The American military, reflecting the attitude of American society, has placed particular emphasis on the concept of social responsibility—the idea that professional officers must use their expertise only for society's benefit" (16). In certain functional areas with an evolution of the military-industrial complex, PMCs are becoming the experts, they are beginning to lobby Congress, and they may one day espouse differing advice from the military. Finally, as Avant (2005) considers, the increasing use of PMCs "may also call into question the esteem with which Americans view the military, changing the value placed on military service ... " (138).

Kant (1983) argued, "if the consent of the citizens is required in order to decide that war should be declared ... nothing is more natural than they would be very cautious in commencing such a poor game, decreeing for themselves all the calamities of war" (113). War is a serious matter, and public debate is essential to ensure states enter war "cautiously." That is one reason for all voters to voice their opinion, and why the Abrams Doctrine sought to make the US government's deployment of forces contingent on the calling-up of the Reserves and National Guard, keeping the ad bellum debate public and the costs of war felt throughout the country. PMCs alter the face of this debate.

PMCs are now the experts, not the military

PMCs are not experts in all aspects of the military, but they certainly have developed resident expertise in many of the areas that traditionally have been the purview of the military alone. This phenomenon is due in part to the actual outsourcing of specific skills. I have already discussed how logistic support is privatized and how eventually the expertise will migrate to private hands. Similarly, war itself is becoming increasingly technological. PMCs seem to manage adaptation to "off-the-shelf technologies" and developing technology quicker than the military itself can grow their own expertise. Look at the privatization of unmanned aircraft because of the sharp increase in the use of and demand of those skills (Mayer 2009; Isenberg 2015).

Without developing expertise internally, formed through experience with the employment of these technologies, the military will be increasingly relying upon PMCs for not only the repair and maintenance but also the "how to's" of employment. A similar privatization is happening in the security service industry. Academi offers merchant mariners training on resisting pirates, and they have also offered anti-piracy forces to deploy to the Somali area (Pitney and Levin 2013; Prince 2014). As the military increasingly turns to PSCs for security operations such as these, even the security of convoys, personnel, or important sites, the experience employing these forces and conducting these operations will reside in the private companies. As Christopher (1999) writes, "It would be ludicrous to permit persons from outside of the profession to make technical decisions regarding how force should be managed in training or on the battlefield" (210). Certainly, at least in the United States, the final decisions are made by the military. However, as the PMCs become more efficient, the military will begin to rely upon them more. So too will the political leadership, especially in areas where the PMCs clearly have developed proficiency. In addition, in the areas where both military personnel and PMC contractors perform the same tasks, there are bound to be conflicts or at least tensions ranging from legal responsibilities and accountability to how the missions themselves are conducted. Some of these tasks, I believe, such as security or interrogation, should not be delegated based on financial savings alone.

Furthermore, not only do PMCs hire away experts at the lower, soldier levels. Large companies like Academi, MPRI, and G4S are hiring away talent from the senior level public sector organizations. In 2005, Blackwater hired Rob Richer, former head of the CIA's near East division and associate deputy director of operations (Isenberg 2009, 64); Enrique "Ric" Prado, who served as chief of operations for the CIA's Counterterrorism Center (CTC), and J. Cofer Black, former head of CTC, also joined Blackwater (Ciralsky 2010).[31] Like retired general military officers, these hirings are coups for PMCs, bringing personal networks and influence, along with their expertise.

The PSC lobby

Whether welcome or not, lobbyists are participants in public discourse. They have some influence in policy development, as the ongoing debate over US Health Care

System changes shows. Speculation on how much influence private companies and lobbyists have is somewhat uncertain; however, millions of dollars are spent by them, generating a fear of vote buying, opacity, and undue influence in the public. The alleged collusion between the defense industry and the Pentagon led President Eisenhower to offer warnings against the rise of the military-industrial complex. Regardless of whether his fears came true, it is impossible to deny the influence of the defense industry on public policy. As the defense industry expands, adding military services, the PMCs become policy influencers themselves.

For example, as Singer (2003) writes, "In the case of the "Train and Equip" program in Bosnia, it is believed that the military consultant firm [MPRI] used its position of trusted expertise continuously to identify additional contract needs for the client" (153). Furthermore, having retired general officers on the boards of PMCs lends credibility with the public, the government, and the military as well. Hartle (2004) writes,

> The military performs its social function through a prescribed set of formal relationships and a vast, intricate set of informal relationships involving both governmental leadership (from the local to the national level) and the business community. As a corporate entity, the military is one of the largest consumers in American society.
>
> (21)

PMCs rely upon many of the same relationships, generating potential conflicts, ones that are now primarily regulated by market forces.

The legislative branch (whether Congress or Parliament), Verkuil (2007) argues, "suffers an information asymmetry vis-à-vis the president and the agencies. Contractors exacerbate this deficiency [by working] outside the usual channels of review and oversight" (192). In Great Britain, this concern led to the release of the Green Paper in February 2002, which looked at different approaches to assert governmental control over British based PMCs. Among the proposals were a licensing regime for PMCs and a system of regulation (Green Paper). There have been some attempts at creating or reinforcing contracting licensing regimes. There is the International Traffic Arms Regulations (ITAR) of 1998, and there seems to be a "stricter interpretation of so-called organizational conflict of interest provisions in the Federal Acquisition Regulation[, which] bar companies building high-tech systems for the military from advising the government on how to assess and manage program progress" (Thompson 2013).

At first glance, a licensing regime looks like it might work. As Leander (2007) writes, "There is a well-established tradition of regulating the role of specialists on violence in this way. However, existing regulation is largely inapplicable to PMCs because it assumes that the relevant specialists on violence are members of public armed forces" (54). Nevertheless, if laws on regulating force were changed to account for PMCs, then it may not work because PMCs, like other industries, have different political channels to pursue than the military.

Finally, even if successful, a PMC licensing regime creates ways for these companies to influence the regulatory standards. Thus, the licensing regime may not be the complete answer. Carafano (2008) offers that

> In the end, the best bulwark against private security abuse in public wars is having members of the private sector who are citizens of responsible states. Sustaining the character of citizenship is a single and most important guarantor that Eisenhower's military-industrial complex will not become reality.
>
> (160–2)

However, the solution assumes that citizenship ideals will trump profit motivation and that home state-based PMCs will only (1) hire its state's citizens and (2) work exclusively for the home state. Let us assume, however, that both a combination of the licensing regime and citizenship-inspired character will work. A different difficulty may still arise.

Competing views

Who best speaks for the military? The answer to the simple question is more complex than a Truman-inspired "the buck stops here" answer. On the one hand, the military itself, through the Joint Chiefs of Staff, advise the political leaders. As Cook (2004) writes, "Military professionals, more than any other group involved in those decisions, possess the professional training and expertise to assess the suitability of the military means under their control for the successful achievement of the politically defined ends" (85). In addition, the defense industry and PMCs themselves have a voice in military matters, generated through their military and political contacts. However, it must be understood that while PMCs are beholden to the government through contracts, they are also responsible to their shareholders. When these interests differ, who wins out? Or, rather, who now has the greater influence?

One indication of the answer is the increasingly acceptable practice of outsourcing seemingly mundane managerial tasks. Nevertheless, when taken as a whole, there is a sense in which the impact of privatization is magnified, and this raises concerns over who or which agency is truly responsible. At times, we believe that bureaucracies seem to have lies of their own; should these be public lives or corporate ones?

As PMCs gain expertise, relinquished by the military, they develop a greater lobbying capacity, and extend their influence through other management services, where military and public policies overlap, such as in counterinsurgency strategy, PMCs will gain a larger role in the ad bellum debate. PMCs, as other private companies, "treat lobbying and advertisement as normal and accepted business strategy and naturally promote their views with administrators, policymakers, and the public at large" (Leander 2007, 57). Undoubtedly, PMCs have become a new player in the decision to go to war. How this debate will transform remains to be seen.

PMCs and peace operations

> Some have even suggested that private security firms, like the one which recently helped restore the elected President to power in Sierra Leone, might play a role in providing the United Nations with the rapid reaction capacity it needs. When we had need of skilled soldiers to separate fighters from refugees in the Rwandan refugee camps in Goma, I even considered the possibility of engaging a private firm. But the world may not be ready to privatize peace.
>
> (Annan 1998)

I have thus far explained how the private military companies affect the military profession and have discussed the entrance of PMCs into the ad bellum debate. In this section, I return to the option of using private forces for peace operations (or as the PMC industry calls them, "Stability Operations").[32] Anticipating a gap in international means or will to conduct these operations, including delivery of needed humanitarian aid, the private military industry has moved into this lucrative business area. The skills needed for these types of operations are not necessarily military, but certainly security of these kinds of operations is one function that is. Furthermore, the large humanitarian requirements are usually only rapidly provided through over the horizon distribution systems—which are found in state militaries. These reasons caused Dag Hammarskjöld to utter his famous claim that only soldiers can perform peacekeeping.

PMCs now argue that perhaps Hammarskjöld was shortsighted—PMCs can be peacekeepers too. When another former UN Secretary General, Kofi Anan, admitted to considering the hiring of PMCs, he was somewhat echoing an idea put forth by Alvin and Heidi Toffler in 1995. In *War and Antiwar*, the Tofflers argue, among other things, that we ought to "consider creating volunteer mercenary forces organized by private corporations to fight wars on a contract-fee basis for the United Nations—the condottieri of yesterday armed with some of the weapons, including non-lethal weapons, of tomorrow" (281). When in 2006, Blackwater offered a brigade sized peacekeeping force to deploy to Darfur, the PMC's offer raised some eyebrows—both within and external to the PMC industry. Erik Prince, founder of Blackwater, also "entertained the idea of building a pre-positioning ship—complete with security personnel, doctors, helicopters, medicine, food, and fuel—and stationing it off the coast of Africa to provide 'relief with teeth' to the continent's trouble spots or to curb piracy off Somalia" (Ciralsky 2010). An often overlooked point is that this brigade could be hired by the UN or other states, in addition to the US government, which Prince admits he had considered.

While peace operations are not the traditional role of military forces, the changing international political landscape has made it so. Combined with requirements for military training, equipment, and organization, militaries in several western states have developed their peace operations capabilities and have included them as part of their core competencies or missions. While hiring PMCs for peace operations may allow the military to focus elsewhere, (1) these PMCs would then have the capability (like the military) to conduct more offensive oriented contracts, (2) their employment changes the security dynamic as well as civil-military relations in areas where they operate, and (3) the PMCs become the market competitors of state militaries, and even munificent PMCs are going to follow the most financially beneficial contracts, perhaps leaving other humanitarian needs uncovered.

I think that the Just War Theory framework also can serve to morally determine when military forces can be deployed to conduct humanitarian operations.[33] I will use the ad bellum tenets, specifically Last Resort, Legitimate Authority, and Public Declaration, to show that PMC forces may not be suitable for peace operations, and the hiring of these forces for these missions and the raising and equipping of these forces encroaches on the military's responsibility, affecting its professional identity.

Using jus ad bellum for peace operations[34]

It seems reasonable that it would be morally permissible to intervene in extreme humanitarian cases, or in cases where vital national interests are at stake. Thus, in order to justify an intervention, some type of convention must be made. For example, one might formulate a set of rules, conditions, or circumstances that would have to be satisfied in order to intervene. (Regardless of which convention is adopted, guidelines for intervention would need to account for the issue of sovereignty—a topic I cover in my *The Problem of Intervention* (1999).)

Instead of inventing some new rules governing intervention, I suggest that we use the tenets of the Just War Theory to morally justify using forces for peace operations. Using the Just War Theory has several advantages. It is well documented, socially, theologically, and legally acceptable, and it has been successfully used to address a number of justified sovereignty infractions. The tenets of the Just War Theory are not uncontroversial, nor are they international law. However, these tenets can provide a moral framework from which the international community can evaluate the situation and inform international legal debate. The international community can then apply IHL to determine whether a state has forfeited its right to sovereignty, when the UN can set aside Article 2(7), and when they can conduct a humanitarian intervention. The question is, "Can the *bellum justum* tenets accommodate the various considerations of peace operations?" I think that the answer is "yes."

With slight modifications, the same war tenets may be used for peaceful intentions, such as intervention for humanitarian reasons. Foremost, an intervention must have a just cause. A state can declare war only if the cause is just. Similarly, the international community must show just cause when it resorts to intervention in a sovereign state. On the surface, it would seem that one could easily fulfill the condition of just cause for interventions. Historically, however, warring states have defined just cause in many different ways. Covell (1998) presents three "traditional" principles for just cause: (1) defense against actual or threatened injury from some other state or states; (2) recovery of or redress for the loss of that which lawfully belonged to or was lawfully due the injured state; or (3) punishment of the state or states guilty of wrong doing. Regardless of what other criteria are used, *bellum justum* theorists commonly accept self-defense as a primary justification for just cause.

Consider the atrocities committed in Bosnia, Cambodia, northern Iraq, and Rwanda. Would not the prevention of genocide and termination of "ethnic cleansing" satisfy the condition of just cause? What about the mass starvation in Somalia? It would seem that the senseless suffering of individuals in those countries would demand intervention. Alternatively, if an operation was intended to keep warring factions apart (peacemaking) or enforce a peace settlement (peacekeeping), then the cause would also be just.[35] Thus, for an intervention to have a just cause, the cause must meet one or more of the following: prevent genocide, prevent ethnic cleansing, prevent other serious human rights violations, or it must be undertaken for the purpose of peacemaking, peacekeeping, and providing a rapid method for distributing humanitarian aid. However, the list is probably not complete. Additional causes that may be defined as just, although they are controversial, include capturing war criminals (e.g. from the former Yugoslavia) and conducting nation building (e.g. in countries like Somalia).[36] There are several situations where humanitarian intervention is justified, such as preventing genocide and peacekeeping. These missions seem to satisfy the principle of just cause.[37] And, applying the tenet of just cause will help ensure interventions are undertaken for the right reasons.

The second condition of jus ad bellum is that the costs of the war must be proportional to the prospective gains. Thus, when fighting a just war the potential gains must balance or outweigh the potential losses. Likewise, the costs must not outweigh the gains. One of the major factors considered before deciding to intervene is the cost of the intervention. Based on the figures from the Defense Budget project, Michael O'Hanlon (1997) estimates the cost for such a humanitarian intervention "might be expected in most cases to range from $3 billion to $8 billion [per year] per 50,000 personnel deployed" (64). The wide range of cost estimates reflects the difficulty in making accurate estimates for any military operation. Terrain, weather, political climate, and even the remoteness of the target state all affect the costs of intervention. Additionally, O'Hanlon's cost analysis does not include the expected loss of life. While accidents occur in any mission, casualty figures for an intervention can vary widely depending on the type of mission, resistance faced, and so on. Although the costs in both material and lives may be high, these interventions are intended to save lives. The problem becomes one of "How much is too much?" What price do we put on human life? Money alone cannot be the deciding factor. An $8 million intervention is small when compared to what some states pay for defense. For example, the 1999 appropriation for US Defense spending was $278.8 billion (*Army Times* 1998, 18). The real issue in any intervention becomes the cost of human life. However, the tenet of proportionality ensures that the international community intervenes only when the gains in lives saved would outweigh the costs of material and lives; and alternatively, would not intervene when those costs and loss of life were too high.

Wars must also have a reasonable chance of success to be considered just, and so must interventions. Those who oppose intervention soon forget that there have been several successful interventions in the last thirty years. Examples include the Indian intervention into what was then East Pakistan (Bangladesh) in 1971, the Tanzanian intervention to stop the depredations of Idi Amin in Uganda in 1978–1979, the

Vietnamese intervention in Kampuchea in 1978–1979, Operation Provide Comfort in Iraq from 1991 on, Operation Restore Hope in Somalia in 1992–1993 (prior to the policy change to pursue Aidid), and Operation Uphold Democracy in Haiti in 1994 (Ramsbotham 1998, 63–4). Several common factors contributed to the success of these interventions. Like any successful business plan or operation, an intervention must be well planned, well organized, and well executed. The social and political environment at the location of the planned intervention must be analyzed and understood *before* the decision to intervene. In addition, contributing states must make available sufficient resources (personnel, material, transportation, and security measures) at the time of intervention. Lastly, "a success objective" (humanitarian end) for the intervention must be defined, so that all participants work toward the same objective.

Any losses incurred during intervention should be proportional to the gains; additionally, the conditions for a reasonable chance of success must be discussed and planned prior to the decision to intervene. Finally, what constitutes last resort may not always be measured in terms of time, because the severity of the situation may dictate the time to intervene. To avoid the questionable intent of covert interventions, the decision to intervene must be made public. Finally, only a legitimate international authority can intervene. I will focus here on the tenets of legitimate authority, last resort, and public declaration because, in each of these areas, the use of PMCs becomes problematic.

Legitimate authority

According to bellum justum, only a legitimate authority can declare war. This condition exists to prevent individuals or small groups that do not represent the state from legally conducting war. However, finding the legitimate authority for humanitarian intervention in the international arena can sometimes become complicated. One might argue that no state should have the legitimate authority to intervene in another, for the simple reason that there would be no system of checks and balances. Recall that this was one of the results from the Treaty of Westphalia. Intervention between states could lead to an unstable pattern of counter-intervention. Unchecked, this could mitigate the humanitarian intents of the original intervention. It is a collective international consensus that helps differentiate between intervention and war.

Fortunately, there is a forum for international consensus. The UN may provide an example of legitimate authority. Each recognized state is a member of the UN, and each member can voice an opinion and vote on UN Resolutions. (Although enforcement is conducted through the Security Council, I will return to this point shortly.) The Organization of American States (OAS) and the Organization for Security and Cooperation in Europe (OSCE) are also examples of potential legitimate authorities that may conduct limited interventions in their areas of influence. (For example, the OSCE monitored the Serbian troop withdrawal from Kosovo and reported the progress to the UN.) If the UN was an international legitimate authority, then the UN could decide when to conduct peace operations. However, the UN has no forces of its own. Therefore, states themselves, as force providers or as lead organizations (or especially when conducting peace operations unilaterally), act as the legitimate authority.

Nevertheless, the lack of resident forces is only one reason that acknowledging the UN as a legitimate authority is not so simple. In his *Morality and Contemporary Warfare* (1999), James Turner Johnson says that one cannot simply designate the UN as a legitimate authority in the same way that states practice legitimate authority, and that any thought of the UN operating as a state would is "based on an imagined reality" (61). The UN, he continues, "lacks in itself the attributes necessary to make it capable of effectively acting out [the] role stipulated in the Charter"—the right to intervene with force, if necessary, to reestablish the peace (61–2). The UN "lacks cohesion," it "lacks sovereignty but depends on agreements among its sovereign member states," and even if member states provide the necessary military forces, the UN does not have an "effective chain of command" (61). And without these, the UN may never have legal or moral ad bellum authority.

While Johnson may be factually correct, the UN has been operating as if it had ad bellum authority, and there is little indication that it stops intervening. Contemporary positive law seems grant the UN authority to intervene, even though it clearly does not match the traditional conception of legitimate authority. So, while this debate on the UN's ad bellum legitimacy should continue, let us assume it for now. How, then, do PMCs change the dynamic? The introduction of PMCs alters the fulfilling of the legitimate authority tenet in two ways. First, within the states themselves, outsourcing can circumvent the normal political decision making processes. For example, in the US, privatization of government functions occurs within the executive branch. As I explored earlier, this outsourcing falls outside of the legislative authority, especially when contracts are for less than $50 million, and we are not discussing the feared foreign policy by proxy. Nevertheless, legitimate authority within the state to hire PMCs for peace operations may look very different from the legitimate authority to send the state's own military forces. (Recall the question of troop caps.) Furthermore, in the United States, the judicial branch could theoretically intervene should the executive branch exceed in the delegation of its authority; that is, if employing PMC forces for normally military tasks is viewed as delegating something inherently governmental. Yet, especially in times of crisis, executive power is seldom restricted judicially. (Consider the Guantanamo controversy.)[38]

Secondly, because the UN Security Council operates on a simple majority, with each permanent member having veto power, there is uncertainty as to which peace operations will be conducted. This problem will not go away with the hiring of PMCs; it might even worsen. However, to many, this problem may not seem so great. They might argue that if the use of PMCs allows one more peace operation than is currently being conducted, then it will be worthwhile. Nevertheless, the same issues that states must consider—including the profit motivation, multiple agents, and so on—must now be evaluated at the international level. For example, what command and control do PMCs fall under for UN peace operations? The UN itself? A regional organization? A lead state? This is not to say that concurrence is impossible or that the benefits to the UN would outweigh other non-financial costs.

Rather, the evaluation of the benefits and costs becomes more complicated, as I discussed in Chapter 4.

Here is another example. What would the international community say to a rich philanthropist's, such as Bill Gates or Richard Branson, proposal to buy an island off of Syria, hire a flotilla, and evacuate refugees, all borne by personal donation?[39] What if the billionaire had to hire some security to ensure the operation was not impinged? If states do not have the will and the UN cannot, it does not seem too farfetched that a rich individual took on such a humanitarian task. Would we feel as comfortable with a Russian oligarch hiring such a force? And, if we find that it might be OK for an individual, then why could we not extend the same reasoning to a group or even a company? If the individual or company hired were sanctioned by the UN, we might have a new reordering of force, but many lives might be saved. I am not advocating for this kind of operation, yet it certainly highlights why hiring private security as a cost saving strategy requires more evaluation. One thing does seem clear: the circumstances must be extreme enough to justify PMC use.

Last resort

War must be a last resort. Similarly, using forces for humanitarian interventions should be a last resort. For example, when the public called for the end of the fighting in Yugoslavia, Security Council Resolutions were ignored, and diplomatic talks stagnated. Interventions such as the NATO air strikes in the former Yugoslavia may be viewed by some as satisfying the tenet of last resort. However, not every case of humanitarian intervention involves a government that is derelict or criminal. In Somalia in the early 1990s for example, there was no unifying, stable government. Thus, some might argue that the amount of human suffering satisfies the tenet of last resort for intervention. Of course, adopting this view generates problems determining when the tenet of last resort is fulfilled, especially when one bases the decision to intervene upon some threshold of human suffering (e.g. numbers of murders or the degree of starvation).

Perhaps even waiting until the last resort could reduce the effectiveness of the peace operation. Williamson (1998) suggests that, "the best time to intervene militarily is early, not for example after sanctions have been tried for considerable time and then adjudged to have failed" (255). However, the last resort for a peace operation need not be *das letzte Mittel* (in terms of time) but rather it should be the *das aeusserste Mittel* (in terms of seriousness) (256). Thus, we should intervene only when the last resort is "most serious." However, this version of serious-based decision making poses problems of its own. Consider the genocide in Rwanda. Recall that the best estimates of the murders committed in Rwanda were between five hundred thousand and one million (Lawyers Committee 1997). Should the intervention have occurred when the casualties numbered in the hundreds of thousands? Common sense would dictate that it is time to intervene. How about when the casualties only reach ten thousand? Intervention would seem a reasonable response. The decision to intervene, however, becomes harder in cases where the numbers of casualties, refugees, or cases of starvation are smaller. One possible way to mitigate this problem of determining when last resort is satisfied is by concurrently looking at the tenet of proportionality. When the expected gains are proportionate to the expected losses, it is time to intervene. In a related way, the decision to use PMCs for peace operations is also tied to considerations of proportionality: Do the benefits of using PMCs outweigh the costs?

The use of PMCs for peace operations may allow for faster delivery of aid, for example, when a state's debate over the deploying the military forces has stalled. While seemingly a benefit, states or the international community can now turn to PMCs and deploy forces when neither the time nor seriousness limit is met. Alternatively, perhaps a seemingly slippery slope move, PMCs could be employed for anticipatory, humanitarian missions—a pre-emptive peace operation. It is also conceivable then that PMCs could be used in a constabulary fashion that could just as well unbalance a delicate political situation (one with a political solution) as it could establish some sense of local, temporary peace. Thus, perhaps unintentionally, PMCs again become an international security entity, competing with states. If states, the UN, or other non-state entities can "use force in a way that reduces or eliminates mobilizing troops, the visibility, sacrifice, and political costs of using force go down," Avant (2005) argues, "If the costs of using force go down, all things being equal, force is more likely to be used. Reducing the process of political mobilization required for action opens way for leaders to take action more readily" (135). And, as I discussed, since PMCs are available to whomever can afford them, then PMCs could make the UN irrelevant in peace operations as well.

Domestic use of force for peace operations raises other concerns. It seems reasonable that in domestic disasters, the last resort tenet still applies—such as when the local law enforcement or first responders are overwhelmed. Although this was arguably the case after Hurricane Katrina, in this example the US government turned to PMCs such as Blackwater to provide security. The lingering question is whether PMCs lower the threshold of last resort. As state and local governments increasingly turn to privatized security, employing forces (at least forces of a certain kind) seems to be becoming less of a last resort and more of a first resort.

Public declaration

> The contractor approach to executive authority hides decisions from public view. Accountability for acts of government is difficult when duties are delegated to private hands and secrecy covers the tracks. Accountability is something that emerges when public acts are challenged or capable of being challenged.
>
> (Verkuil 2007, 13)

A just war and a just humanitarian intervention must be publicly declared, and satisfying the public declaration tenet for intervention is a simple process. Publicly declaring an intervention could merely be a matter of passing a Security Council Resolution (SCR). Some past examples include the passage of SCR 688 for

Operation Provide Comfort in northern Iraq, SCR 792 for the humanitarian intervention in Somalia, and SCR 867 for peace operations in Haiti. Specifying that all justified interventions must be publicly declared also eliminates questionable justification for the "covert" interventions in Africa, Southeast Asia, and Latin America that were so prevalent during the Cold War. It seems clear that states that provide forces for peace operations, either military or PMC-based, should also publicly declare their intentions.

Included in the public declaration, at least for democracies, is a corresponding public debate. However, because PMCs can be employed without legislative oversight, one must consider whether peace operations can be conducted without either the declaration or the debate.[40] It may seem implausible that states can conduct large-scale peace operations without public debate or scrutiny (at least in "open" states); perhaps that would not happen.

Nevertheless, concerns remain of the possibility of small operations being conducted without the debate. The military must answer to Congress; PSCs only have to if subpoenaed. And, in the case of Blackwater/Academi/Xe, the government itself contributed to this air of secrecy. Blackwater's State Department contract, for example, forbade their employees from talking with the media (Prince 2014, 202). This State Department legal clause merely reinforces the practice that exists when using PMCs for training foreign military.

For peace operations or for war, hiring PMCs allows for less public scrutiny. Using PMCs allows for fewer political decision-makers to be involved. Because the political commitment seems lower, there is less incentive to have the public debate —the same debate is critical for upholding the public declaration tenet of the Just War Theory. Highlighting this shift in public debate when PMCs are deployed, one could look at her local paper and get daily updates on the military casualties in Iraq and Afghanistan. But, she would not be able to find similar statistics on PMC contractors killed. More troubling is the lack of public discussion on the care of wounded contractors.[41] One cannot assume then that things would be different using PMCs for peace operations.

Lasting effects?

The hiring of PMCs will not erase the military profession overnight. In fact, it is not all clear what the return to privatized security entails. Nevertheless, to deny that there is no impact would be both naive and imprudent. Even with the 2010 reduction and withdrawal of troops in Iraq, there appears to be an acknowledgement that PMCs are here for the foreseeable future. Bacevich (2008) notes,

> To salve the nation's conscience, the government might augment our hard-pressed troops with pricey contractor-mercenaries, but it won't actually trouble citizens to do anything ... Americans may not like mercenaries, but many of them harbor an even greater dislike for the prospect of sending their loved ones to fight in some godforsaken country on the other side of the world.
>
> (155)

If views such as Bacevich's are correct, then in each of Huntington's three areas of expertise, corporateness, and responsibility, military outsourcing is affecting the military profession.

First, there is competition over the talent and experience that both the military and private contractors rely upon, not to mention a progressive outsourcing of the profession's custodianship over its military expertise. Second, the growing acceptance of hiring market-based employees is a challenge to the military's corporateness, which in turn may alter how a soldier views her own identity within the military profession. Third, the military's advisory responsibility to the civilian governmental authority will alter as privatized security gains an accepted voice within government.

While these changes in no way guarantee a lower threshold for going to war, increased reliance upon—and a building acquiescence of—PMCs alters the jus ad bellum debate landscape. How we think about war and how we decide when to resort to war must be explored anew with the inclusion of these other private actors. Even when deciding to use PMCs for humanitarian action highlights how the debate will change. Wakin wrote in 1979 that "In the hands of the mediocre or the morally insensitive, the vocation of arms could find its noble purpose distorted with tragic consequences for all humanity" (9). One might dismiss his view as too strong; nevertheless, some may argue that the cost of using privatized military outweighs the benefits, so we ought to have a prima facie ban on their use except in extreme circumstances. Until we can fully explore how privatized military will change the military profession, then adopting this kind of policy may be necessary.[42]

Notes

1 Also see Frank Langfitt's "Private Military Firm Pitches Its Services in Darfur" (2006) and Sharon Weinberger's "Facing Backlash, Blackwater Has a New Business Pitch: Peace-keeping" (2007).
2 Cf. US Department of the Army's *Peace Ops:* Field Manual 3-07.31 (2003a), *Stability Operations and Support Operations*, Field Manual 3-07 (2003b), the US Joint Chiefs of Staff. *Joint Tactics, Techniques, and Procedures for Peace Operations* (1999), as well as the US Army Peacekeeping and Stability Operations Institute (PKSOI): https://pksoi.army.mil/.
3 For a fuller discussion of legitimate authority and humanitarian intervention, see David Barnes' *The Problem of Intervention* (1999).
4 Cf. Morris Janowitz's *The Professional Soldier* (1960), Allen R. Millett's *Military Professionalism and Officership in America* (1977), Don M. Snyder, John A. Nagel and Tony Pfaff's *Army Professionalism, The Military Ethic, and Officership in the 21st Century* (1999), and Don M. Snider and Gayle L. Watkins' "The Future of Army Professionalism" (2000).
5 It is interesting that many of the larger PSCs and PMCs have not only developed ethics codes or codes of conduct, but have also publically included them on their websites. For example, see Engility Corporation (http://www.engilitycorp.com/about/ethics/), Triple Canopy (http://www.triplecanopy.com/trusted-partner/ethical-approach/), DynCorp International (http://www.dyn-intl.com/about-di/values-code-of-conduct/), or Aegis World (http://www.aegisworld.com/compliance/). One might argue that these companies are merely making shrewd marketing ploys, but this author likes to think that these codes reflect a strong desire for these companies to remain morally legitimate.

6 For a study of the relationship between the oath of office and a professional military ethic, see Hartle's (2004) Chapter 4, "The American Professional Military Ethic," 42–75.

7 Verkuil (2007) continues, "[Military] officers exercise command authority that cannot devolve to private contractors. Thus, when private contractors interrogate prisoners in Iraq or participate military actions involving the use of force, they usurp public authority, unless Congress has approved. The exercise of this authority is a public function, which makes the phrase "private military" an oxymoron offensive to our Constitution" (103–4).

8 The executive summary continues, "We therefore conclude that contractors should not be deployed as security guards, sentries, or even prison guards within combat areas. [PSCs] should be restricted to appropriate support functions and those geographic areas where the rule of law prevails. In irregular warfare (IW) environments, where civilian cooperation is crucial, this restriction is both ethically and strategically necessary." (Disclosure: The author was both a panel chair at the conference and helped craft this summary.)

9 The threat of a modern-day military coup in the United States is a periodically popular topic of books and film. One example, *Seven Days in May*, 1964, looks at a coup plot where members of the military and others look to overthrow the fictional President Jordan Lyman in seven days.

10 Sandline's chief, Spicer, saw the situation in a different light; see Drohan (2004, 223–4).

11 Avant continues, "Over the long term, if the shape of professionalism changed in a way that reduced the professionalism of the Army, at the extreme perhaps making dying (or killing) for one's country less legitimate, one could imagine such changes eroding the fit between the dimensions of control and decreasing military effectiveness" (120).

12 The US government also looks to private companies to continue to support training activities abroad. In addition to contacted training in Afghanistan, PMCs are used to both train Iraqi forces and Syrian opposition forces, and PSCs are hired to provide security for those training. In January 2016, outsourcing training took an interesting turn. Reporter Tim Shorrock (2016) wrote that General Dynamics Information Technology was contracted for training Iraqi Special Forces, who are fighting ISIS, and that three US contractors who were kidnapped allegedly worked for Sallyport Global, a company that provided PSC services to General Dynamics. According to their website, Sallyport Global denies that these personnel were employees of their company: "Statement Regarding Missing Individuals in Iraq: Sallyport's first concern is for the individuals who were reported missing. The three subcontractors who were reported missing January 16, 2016 in suburban Baghdad are not employees of Sallyport or any affiliated company." (http://www.sallyportglobal.com/).

13 Avant writes, "As is often the case, however, the cost of garnering the exact same service from the private sector was higher. Indeed, from the start, the privatization of ROTC is expected to cost more than the alternative of using active duty personnel. The RAND Corporation estimates suggested that each year it costs about $10,000 more per instructor for the private staffing of ROTC training in the United States. In 2002, the contract for ROTC training was re-competed and MPRI lost the contract on cost concerns. Many worry, though, that the new contractor ... would field less qualified staff. They pointed out that officials from MPRI routinely participate in intellectual forums sponsored by the Army [implying that the competition did not]–partly because MPRI has such an impressive array of retired, high-level staff–and thus MPRI is considered well positioned to do a better job keeping up with the implicit requirements for ROTC instructors and trainers (enthusiasm, experience, professionalism). Paying attention to cost alone, these people argued, would generate a less impressive program" (117–18).

14 Cook continues, "Result of that extensive professional development is, indeed, a degree of intellectual independence from the society they serve, and it is grounded in their unique professional focus" (63). I will return to this point in the next section.

15 The author was a student in the pilot implementation of ILE in 2001.

16 Carafano adds, "The greatest danger about the future of war is the danger of outsourcing imagination to others" (207). Nevertheless, one could point to the large number of private and politically influential think tanks as another accepted practice of outsourcing. This time, however, instead of logistics or security, these entities explore policy, including the future of war. I will turn to these organizations later in the chapter.

17 Janowitz adds, "Monetary rewards might work most effectively for those officers engaged as military technologists. Even if salaries were to become truly competitive, the incentive system would not necessarily produce the required perspectives and professional commitments" (422).

18 For example, "In the Kosovo War, for example, the going rate for professional soldiers to help the rebel KLA group was a reported $4,000 per month," writes Singer (2003, 43).

19 See Janowitz (1960, 420). He writes, "[the military officer] is subject to civilian control, not only because of rule of law and tradition, but also because self-imposed professional standards and meaningful integration with civilian values." Carafano (2008) would counter that "Part of making profits in a transparent, global competitive economy is avoiding scandals that attack the credibility of the company" (167). Unfortunately, past behavior indicates that the market system to regulate PMCs is falling short of expectations.

20 Colonel Bruce Grant of the Institute for National Strategic Studies, writes, "MPRI can send twenty former US colonels to Bosnia, while the US Army would have to strip more than an entire combat division to muster that many" (Silverstein and Burton-Rose 2000, 166–7).

21 Erik Prince makes a point of the behavioral expectations of his former employees. See Prince (2014, 173–89).

22 For a glimpse into the debate surrounding the "proper" roles for retired flag officers, see Don Snider's "The Army's Ethic Suffers under Its Retired Generals" (2009), David Margolick's "The Night of the Generals" (2007), Richard J. Whalen's "Revolt of the Generals" (2006), and Martin Cook, "Revolt of the Generals" (2008).

23 The author worked with several Army mentors. See Tom Vanden Brook, Ken Dilanian and Ray Locker's "Retired military officers cash in as well-paid consultants" (2009).

24 Verkuil (2007) writes, "In seeking leaders for public service, the Volcker Commission praised the high levels of performance in the military. Military leaders are adept at management under conditions of uncertainty and adversity. And they're particularly well trained. The military invests more in leadership education and training that its civilian counterparts. In encouraging these leaders to remain in service rather than enter the private sector, the military should consider whether its increased use of private military contractors itself creates competition for its very best people" (174).

25 Leander (2007) writes, "Most countries have well-defined institutional arenas where the armed forces can be consulted and asked to express their views. Characteristically, these are concentrated with the executive branch of government, which is often more often than not given primacy in deciding on the use of force . . . . The participation of specialists on violence in other words tends to be conceived in strictly and well defined consultative terms" (55).

26 Interestingly, though, Huntington does think that the military's enlisted personnel share this responsibility. He writes, "The enlisted men subordinate to the officer corps are a part of the organizational bureaucracy but not of the professional bureaucracy. The enlisted personnel have neither the intellectual skills nor the professional responsibility of the officer. They are specialists in the application of violence not the management of violence. Their vocation is a trade not a profession" (17–18).

27 For example, GEN Clark faced a similar dilemma during the Kosovo Campaign, Operation Allied Force. See Clark's *Waging Modern War: Bosnia, Kosovo, and the Future of Combat* (2001).

28 The State Department's Policy Planning Staff provides a listing of these organizations, which numbers more than 50, including RAND, The Brookings Institution, The Cato Institute, American Enterprise Institute, International Institute for Strategic Studies, Heritage Foundation, Center for Strategic and International Studies, Center for a New American Security, and the Council on Foreign Relations, among many others. See http://www.state.gov/s/p/tt/, accessed 14 December 2009.

29 Singer writes, "[PMCs] have begun to be utilized as an alternative way to circumvent these policy restrictions. The intent of privatized military assistance is to bypass Congressional oversight and provide political cover to the White House if something goes wrong. Beginning in the second term of the Clinton administration, the United States quietly arranged for the hire of a slew of [PMCs], whose operations in Columbia range far beyond the narrow restrictions placed on US soldiers fighting the drug war" (2003, 207).

30 Carafano (2008) writes, "Covertly contracting foreign policy is one of the least likely avenues for advancing a covert administration agenda. Government contracts are subject to the full gamut of government and public oversight. The administration must, for example, comply with the requirements of FARS, have money appropriate by the Congress, make its practices opened to oversight by Inspectors General, government auditors, the Government Accountability Office, and Congressional committees, and comply with "sunshine laws" such as the Freedom of Information Act … " (162).

31 Likewise, according to Shorrock (2016), Sallyport Global's parent company DC Capital Partners "are advised by a star-studded list of former Pentagon, intelligence, and State Department officials who should know something about security," including Michael Hayden, Richard Armitage, and Anthony Zinni.

32 By "peace operations," I refer to "a huge range of connected military, diplomatic and humanitarian tasks, as diverse as reforming justice and security systems, disarming and demobilizing troops, reintegrating them into peaceful pursuits, and supporting humanitarian assistance." See Canada's Foreign Affairs and International Trade Canada (DFAIT).

33 I argue for the adoption of a modified Just War Theory framework for peace operations in my *The Problem of Intervention* (1999).

34 Another theorist, James Pattison, has also looked at using private forces for peace operations. See his "Outsourcing the Responsibility to Protect" (2010b) and (2014).

35 For a further discussion see David Fisher's "Some Corner of a Foreign Field" (1998).

36 See also Comfort Ergo and Suzanne Long "Cases and Criteria: the UN in Iraq, Bosnia, and Somalia" (1998).

37 Orend (2013, 94–103) offers an additional, humanitarian casus belli to complement the Just War Theory's defense against aggression just cause for going to war.

38 Verkuil (2007) writes, "In a military setting, the president is overseeing unprecedented delegations of combat authority to private hands in situations where Congress has not acted to authorize such delegations … .. To make matters worse, privatized actions are often nontransparent. FOIA is inapplicable to private contractors, a deficiency of democratic control. The desire for secrecy may be one of the motivations for executive delegations of significant authority to private contractors, at least for some presidents. During Iran-Contra, the Reagan White House in effect ran a private war in Nicaragua against Congress's instructions" (105).

39 An example like this was brought up by my students in class.

40 Percy (2007b) writes, "The state might also find it easier to sustain wars that might go against public opinion because of the presence of PSCs. While pre-nineteenth century commentators worried that the use of mercenaries would lead to domestic tyranny, today the concern is that private force could lead to a tyrannical foreign policy detached from popular control" (237). Although Percy here is worried about using PMCs to conduct more belligerent activities, the same concern applies to conducting peace operations "detached from popular control."

41 See T. Christian Miller's "Sometimes it's Not Your War but You Sacrifice Anyway" (2009). I will return to the problem of contractor casualties in the next chapter.
42 I would like to thank members of the audience, especially Christof Tatschl, Max Borne-feld-Ettmann, David Kieran, and Tânia Pinc, at the 2010 International Studies Association (ISA), February 2010, New Orleans, for their comments on an earlier version of this chapter.

7 The "second contractor war" and the future of armed contractors

Whatever Blackwater's motives, I won't join the "moral giants" who would rather do nothing at all than send mercenaries to Darfur. If the Comintern could field an army and stop the killing, that would be all right with me, too. But we should acknowledge that making this exception would also be a radical indictment of the states that could do what has to be done and, instead, do nothing at all. There should always be public accountability for military action— and sometimes for military inaction as well.

(Walzer 2008)

This book so far has attempted to address some of the fundamental ethical issues associated with the hiring and employment of armed private contractors. In order to understand the phenomenon itself, we spent some time explaining both the historical lineage of private force and how it grew and waned with the treaty of Westphalia and the rise of the nation state. We discussed this private military rebirth: how it is similar to its mercenary descendants and how it is very much different. Perhaps the return to private forces by states and others is truly an evolution of force as we know it, or perhaps governmental outsourcing efforts, combined with the post 9/11 wars in Iraq and Afghanistan, formed an ideal nexus where private force emerged and will gradually disappear over time.

While I do think that geopolitical conditions, operations in far-reaching areas, as well as Western governments' attempts at outsourcing did create the environment where private security companies thrived, as another author put it, this was the First Contractor War (Hagedorn 2014).[1] Based upon the emerging uses of private forces combined with an acceptance by the United States, Britain, and other states that PMCs and PSCs are part of the way states do business, it seems that the private security company is here to stay.

So, what is the status of armed private contractors today? Recall that in 2010, President Obama declared the end of the Iraq war, and much of that fight and the contractor support of that operation seemed to wind down. After combat troops left Iraq, logistics PMCs continued to close bases there. Of course, the use of privatization by the United States did continue. Much of the United States current involvement in the Middle East and elsewhere is carried out by privatized military companies. Furthermore, other states and non-state actors are turning to privatized security solutions. Therefore, while there is a strong argument against marketing force, while there is no clear delineation morally or legally between public and private forces, and while the long term impacts of increased reliance on privatized military remains uncertain, it is worthwhile to look at the current use of privatized military and the future proposals of private force, then layout some recommendations to better understand where this industry is going, how to harness it, and how to ensure that moral, and not just the political and legal, considerations are debated for the use of PSCs in the future.

The second contractor war

Ann Hagedorn (2014) astutely asks whether we are entering a Second Contractor War, and I think her hypothesis is both accurate and telling. As of summer 2015, there were 3,550 US troops in Iraq (Baker, Cooper, and Gordon 2015), and there is a growing concern of "creeping instrumentalism," as the strategy for defeating ISIS evolves. They are being supported by 1,349 DOD contractors (Contractor Support 2015) and upwards of 5,000 DOS and other contractors (McLeary 2015b). There are still about 42,000 DOD contractors working in the CENTCOM area of operations, and while the focus on the use of private military is centered on the Middle East, contractors are used worldwide. Kate Brannen (2015) reports that SOS International (SOSi) was awarded a $400 million contract for work in Iraq, including "a $40 million contract to provide everything from meals to perimeter security to emergency fire and medical services at Iraq's Besmaya Compound, one of the sites where US troops are training Iraqi soldiers." And, SOSi has a $100 million contract at Camp Taji for the same kind of support.

What is more interesting is the SOSi $700,000 contract to provide "a small group of security assistance mentors and advisers" to Iraqi Ministry of Defense and Kurdish officials that could increase to $3.7 million (Brannen 2015), as former, high ranking US officers are part of this advisory plan. Recall, that the Obama administration also asked Congress for $500 million to train and equip moderate Syrian Rebels, and much of this planning includes how private military companies will be employed in support of this mission, especially if the "no boots" constraint remains. Even if US policy changes and ground forces are deployed to Syria or Iraq, private contractors will be sent to support them as well. As one former private contractor said, "this new war will present an opportunity for the companies that have a resident train and advising capability to contribute to this new effort" (Lake 2014). The $500 million President Obama asked Congress to authorize to train a new Syrian opposition out of Saudi Arabia is part of a $5 billion fund Obama requested in spring 2015 from Congress to help train and equip US allies to fight terrorists.

Nevertheless, the early stages of the use of Syrian fighters has not gone exceptionally well. McLeary (2015a; Schmitt and Hubard 2015) wrote that the first 50–60 man New Syrian Army contingent (part of the so-called Division 30 program) was attacked by al-Nusra Front when they entered Syria. Also, the entrance of overt Russian military support is complicating efforts. Since the Russians back the Assad regime, an increased Russian military presence could result in potential direct confrontation. If the Russians become more of participants in the Syrian conflict than just advising, then the situation could become far graver.

Others are looking at ways to privatize the war in Syria. Lim (2014) offers that privatizing the Syrian war could save lives. While it seems unclear exactly how a private military peacekeeping force would end the conflict with fewer casualties, certainly a case could be made for a UN-sanctioned and purchased force to help stabilize the peace, assuming that all sides agree to it. This kind of UN privatized force is in keeping with the peacekeeping force Walzer (2008) describes and cautiously accepts. Erik Prince, on the other hand, thinks that private contractors are the solution to actually end the fighting. He believes that they are willing and best positioned to join the fight (Rosen 2014; Lamothe 2014). Prince said:

> If the old Blackwater team were still together, I have high confidence that a multi-brigade-size unit of veteran American contractors or a multinational force could be rapidly assembled and deployed to be that necessary ground combat team. The professionals would be hired for their combat skills in armor, artillery, small unit tactics, special operations, logistics, and whatever else may be needed. A competent professional force of volunteers would serve as the pointy end of the spear and would serve to strengthen friendly but skittish indigenous forces.
>
> (Drennan 2014)

Prince's proposal harkens back to the days of EO and Sandline, where Operational Support PMCs actually fought as combat units. Additionally, Nissenbaum (2014) reported, "A group led by a former Pentagon official devised a plan to supply moderate Syrian rebels with weapons sourced in Eastern Europe and financed by a wealthy Saudi." With no clear end in sight for the complex Syrian conflict or the ISIS fight, the opportunities for private contractors will continue. Former ISOA president Doug Brooks relates, "these are shoes instead of boots on the ground" (Lake 2014).

Piracy and Private Maritime Security Companies (PMSCs)

The demand for private security has not only continued on land but the increased need for security to protect shipping around the globe has also generated a growing maritime private security business. Increased piracy off of the coast of Somalia resulted in an increase in PSC-led training, and it offered employment for private contractors on the seas (Houreld 2008). Erik Prince (2014) discusses how Blackwater used its training facilities in Moyock, NC to train sailors in force protection and counter piracy measures, which grew out of their Navy contract to "train sailors to identify threats, engage enemies, and defeat attacks underway in ships in port and at sea" (45), after the USS Cole attack. Blackwater even created a Maritime Security Solutions company, which refitted a ship, the McArthur, to assist with antipiracy operations in the Gulf of Aden (93).

Somalia was not the only area to see a rise in private maritime security companies (PMSCs). The Anambas Islands and Singapore Strait also has its share of piracy. Cheney-Peters (2014) notes that the insurance firm Allianz reported a "700 percent rise in actual and attempted attacks occurring in Indonesian waters in a 5-year span, from 15 in 2009 to 106 in 2013" (5). While states' navies patrol these waters, it may be cheaper, especially in the short run, to contract PMSCs to help protect shipping, and some of the weaker states can turn to contracting maritime security easier than building the capacity in their nascent naval forces.[2]

Yet, private security on the high seas has its own challenges. As discussed in Chapter 2, defining piracy continues to remain unclear in the international law arena. The international laws are different for territorial waters and the open sea. One problem left to be sorted is the legal status of pirates when they operate close to shore; are they still pirates? Furthermore, this distinction also affects those providing security. As Cheney-Peters (2014) writes, when "operating in territorial waters, the UN convention on the Law of the Sea (UNCLOS) provides little clarity on the legal status or protections for PMSCs" (8; Pitney and Levin 2013, 107–8). It is illegal in some ports to bring arms into port onboard the ship. Yet, many of the PMSC contracts are so lucrative that some companies merely dump their weapons overboard before entering these ports, creating a cottage industry of floating armories to support these operations (Cheney-Peters 2014, 9). And, while there have not been reported incidents where PMSC contractors have killed civilians, like the Nasoor Square incident, there was a legally complicated event where two Italian Marines on Vessel Protection Detachment (VPD) duty fired upon and killed two Indian fisherman, whom they believed to be pirates (Pitney and Levin 2013, 114). A legal fight over jurisdiction followed, and these legal concerns are only magnified when security contractors are providing the VPD service instead of a state's military forces.

Other countries' PSCs

The United States and her allies are not the only private military players now; Russia, Israel–even China have entered the market for force. There seems to be a growing number of Russian PSCs, especially as laws forbidding PMCs in Russia seem to be changing. In a confusing and ill-fated incident, it seems that in 2013, "the Moran Security Group, a Moscow-based private military company, . . . contracted a Hong Kong-registered entity called Slavonic Corps Ltd. to dispatch armed personnel to Syria" (Weiss 2013). At that time, PMCs were essentially illegal in Russia. When these contractors returned home, they were arrested and prosecuted for violating these laws. Furthermore, the continuing conflict in the Ukraine introduced new Russian supported forces. Josh Rogin (2014) reported that "Private security contractors working for the Russian military are the unmarked troops who . . . seized control over two airports in the Ukrainian province of Crimea, according to informed sources in the region." While it is not yet clear whether these forces were Russian security contractors, Russian forces in disguise, or proxy forces, the lack of uniforms combined with the fighters' abilities and equipment led some like Rogin to theorize that these were Russian PSCs. Interestingly, these alleged employments of Russian contractors was occurring while a debate over the legality of Russian PMCs was occurring in the Duma ("Bill to Allow" 2014). As the laws regulating PMCs change in Russia, companies like Russian PSC, Russian Security Systems (RSB) Group (rsb-group.org), are advertising openly over the Internet According to its website, RSB offers a full "complex of services concerning armed protection and security provision outside Russian Federation."[3]

In addition, China, looking to secure its own citizens overseas and to enter the lucrative private security industry, has expanded its own PMCs. Erik Prince's own Frontier Services Group has an article highlighting this Chinese expansion. The analysis article notes that "Chinese privately-owned security companies, both at home and abroad, were only made legal under Chinese law in 2010 ... [and] a dozen of them are now testing the water in Africa and the Middle East." Companies like Dewei Security Group Ltd. and Shandong Huawei Security Group are already employing armed guards, and Huaxin China Security is "already providing armed Chinese guards to escort cargo ships sailing to the Horn of Africa." As Erickson and Collins reported in 2012, China has "at least 847,000 Chinese citizen workers and 16,000 companies scattered around the globe," and they will need security.

The future of armed military privatization

While history suggests that humans are notoriously bad at predicting the future of global events, the recent past activities of private military sheds some light on what we might expect in the coming years. Certainly, the threat of piracy will not go away soon, nor does it look like there is a simple, quick solution to the Syrian conflict. As discussed, other countries are flexing their private military muscle. So, what else can we look forward to seeing in the PSC realm? The future of military privatization may include, as Prince suggested, a reappearance of Operational Support companies like EO and Sandline. While many assume that the days of these kinds of companies, ones actively engaged in offensive operations are past, it would be wise to remember the Wonga Coup. As recent as 2004, while PSCs were growing and operating in Iraq, Margaret Thatcher's son, Sir Mark Thatcher, was involved in a coup attempt in Equatorial Guinea. With a plot like the Forsyth novel, *The Dogs of War* (1974), Thatcher, Simon Mann, and others, 70 of whom were arrested as mercenaries in Zimbabwe, hatched a plan to overthrow Teodoro Obiang Nguema Mbasogo, leader of Equatorial Guinea (ironically who took over in a 1979 coup) (Wines 2004; Roberts 2006). Thatcher was not the head of a PSC; clearly this was a group of mercenaries. But, this incident highlights that there are those who will turn to private force to solve their political problems. There remains, in an uncertain world, an apparent need for private force that some future PMC could once again fill.

These sort of plots do seem farfetched, but there is a noticed increase in the use of foreign fighters in places like Syria, and this may make a turn back towards the employment, if not the acceptance, of both less organized mercenary companies like Slavonic, to other security companies filling a void in force. We have heard that many foreigners are joining both ISIS and the fight against them. Reuter (2015) reports that there are Afghans fighting for Assad in Syria, as Iran and Hezbollah seek to replenish their losses, this time with refugees from the war in Afghanistan. Do we consider groups like Slavonic PMCs, or are they mercenary groups? It is not so clear.

Furthermore, Adam Nossiter (2015) reported that "Hundreds of mercenaries from South Africa and other countries are playing a decisive role in Nigeria's military campaign against Boko Haram," and Nigerian Rear Adm. Okoi said that South African contractors were hired to train Nigerian troops. Roston (2015) also claimed that Erik Prince made a business proposal to the president of Nigeria to use contractors to fight Boko Haram. Rostom added, "It is not clear that the South African mercenaries now reportedly operating in Nigeria are in fact doing the job that Prince had bid to do or even whether they are employed by Prince's company." As the need for forces and expertise continues in Nigeria and other conflict areas, there is a corresponding gap for private military solutions, and the private security industry will evolve to satisfy these needs.

Therefore, a look at the history of Blackwater, now Academi, and its founder Prince's Frontier Services Group, might illuminate the possibilities for future PSCs. While Blackwater was chiefly known for their security services in Iraq, they, according to Prince (2014) and others, have been involved in a variety of security operations. Apart from the security training, antipiracy assistance, and protective details, Blackwater (Xe/Academi) had its own airwing, Presidential Airways, and its business expanded to include other affiliates and subsidiaries that spanned the private military services, including Backup Training Corporation, Pelagian Maritime, Greystone, Constellation Consulting Group, GSD Manufacturing, and Raven Development Group (Prince 2014, 293). As late as 2009, Xe maintained and equipped the CIA's drones in Afghanistan and Pakistan ("CIA cancels" 2009). While, as Isenberg (2015) notes, contractors were not the ones dropping ordinance, they were involved in all support activities before and after the in-flight strikes.

But, Blackwater was also allegedly participating in more sensitive operations. Scahill (2009) wrote in *The Nation* that Prince told "*Vanity Fair* that Blackwater works with US Special Forces in identifying targets and planning missions, citing an operation in Syria." And, he noted that the company was working with Joint Special Operations Command (JSOC) and the CIA. Blackwater Select allegedly was the operations element and Total Intelligence Solutions (TIS) provided intelligence analysis. Scahill also offers that some Blackwater personnel worked with the Pakistani Frontier Corps. Prince himself has taken his lessons learned from running Blackwater/Xe/Academi and its affiliates with him, and he has formed a new company: Frontier Services Group. According to Roston (2015), Prince's company, Frontier Services Group, incorporated in Bermuda, is "a company traded on the Hong Kong stock exchange. The firm is backed by the CITIC Group, which is majority-owned by the government of China."

As alarming as these supposed activities seem, it is clear that there remains a niche for these kinds of expertise, expertise that can only be developed after years working with and for special operations forces like JSOC. While there is concern about the transparency of such contracts, as well as the question of international law, there also remains an overarching concern: where should the privatization line lie? Perhaps we need not look too far into the future to see what private security will become; the future is now. As Wilkerson, Colin Powell's chief of staff (2002–2005) said, "It wouldn't surprise me because we've outsourced nearly everything" (Scahill 2009).

The problems remain (and what to do about it)

One thorny issue that still surrounds the hiring of PSCs is legal jurisdiction. Legally US PSCs fall under legal jurisdiction through the Military Extraterritorial Jurisdiction Act (MEJA) and UCMJ—with words added in the 2007 National Defense Authorization Act (NDAA) to expand those covered to "persons serving with or accompanying and armed force in the field." Furthermore, "war" was replaced by "declared war or contingency operations." In 2009, two Blackwater subcontractors were charged under MEJA for a shooting incident in Kabul (Prince 2014, 289). So far so good for refining the legal authority over PSCs working with the US military. But, these may not include non-DOD PSCs. US non-DOD PSCs fall under the US Code 18 Section 7: Special and Maritime Territorial Jurisdiction of the United States. A CIA contractor in Afghanistan, David Passaro, was convicted of assault of a detainee under this jurisdiction (Jones 2012). Nevertheless, not all contractors are hired by the US government; many, as we have discussed, are hired not just by other states but through other PMCs, multinational corporations, and NGOs.

Yet, even with DOD PSCs, there are other issues that must be explored. It is one thing to say that UCMJ now provides the DOD legal authority to prosecute crimes committed by PSC personnel—even in bello violations. Nevertheless, it is quite a different puzzle as to whether the DOD PSC contractor must follow the legal orders of superior officers, as is legally required for military personnel. Certainly, this kind of legal, military authority is outside of normal security contracts. It seems like we may have a case of the military having legal responsibility but not the same level of legal authority over PSC personnel. While this example may seem nitpicking, there is the famous case, discussed in Chapter 5, of military personnel taking orders from Blackwater contractors at the CPA headquarters in al Najaf in April 2004 (Prince 2014, 133–47).

What if states licensed PSCs for their security services? Some might propose that states deputize the PSCs. While many on both sides of the PSC debate argue for better regulation, and certainly a licensing regime could increase the contractual control and oversight of employing PSCs, the argument itself does not change. Currently, an export license by the State Department's Office of Defense Transitions Assistance is sufficient for PSCs that are based in the United States; these are not hard to come by. This example highlights the weak end of a licensing spectrum. At the other end, one might suggest making the PSCs part of the military. This option seems to solve many of the control issues, but also removes the services that the PSCs provide from the market that was not only supposed to regulate the PSCs, but more importantly, provide the cost savings and increased efficiency that caused the state to look towards outsourcing in the beginning. A licensing regime would need to fall somewhere in the middle, and its effects would have to be examined to determine if the benefits would outweigh the costs.

Another underexplored, and underreported, area that must be brought into the debate over the use of private force is contractor casualties. Unlike military casualties that were forefront in the news during Iraq and Afghanistan, little notice was given to the sacrifices of the private contractors who shared the battlefield with the military personnel. Speaking at Arlington national Cemetery, President Obama commented on the sacrifice of the more than 2,200 service members who have died in Afghanistan. What's overlooked is that since 9/11, 1,592 contractors have also died in Afghanistan. As of May 2015, 101 contractors have died there (Zenko 2015). According to the Watson Institute for International and Public Affairs, 3,481 US contractors died in Iraq between March 2003 and April 2015 ("Costs" 2015), and "Foreign workers for US contracting firms often do not have their deaths recorded or compensated." While the military monitors injuries, woundings, and deaths of its personnel, there is no systematic government program to do the same for contractors. Furthermore, unless the contract stipulates it, the families of killed contractors do not receive death benefits like those military families do. In this way, the human costs of contracting are often hidden from public scrutiny, which further exacerbates the perception that the use of PMCs is secretive.[4] Nevertheless, the problem is more than mere transparency and a matter of public opinion. Security contractors are only owed medical support based on what is stipulated in their contracts. As a soldier, I and my troops understood that our mandate to "leave no man behind" included those US contractors that worked with us, but this was more from a sense of doing what was right as opposed to doing what was legally mandated or expected. My medics treated contractors and civilians as well as they treated other soldiers. But, that was in the fight. What about contactor care after the fight—when the contract is complete?

The US military tracks the number of deployments each person makes, but there is no capturing of this data for contractors, except perhaps by the private company. Additionally, there are programs, such as the Army's Warrior Transition Units, that are designed to assist the soldier through her care and transition back to civilian life. There is no equivalent program for military contractors, unless their respective companies have a program of their own. And, there is no program to track and help with contractor post-traumatic stress disorder (PTSD). Finkel (2013) notes that between 20 and 30 percent of the approximately two million US service members who served in Iraq and Afghanistan have returned home with PTSD. Can we accurately extrapolate the numbers of contractors who are suffering the same effects? Even if we can, this problem is hidden from view, and these contractors may not be getting the care they deserve.

Therefore, one of the major initiatives must be gaining thorough accountability of the private military contractors whom states are hiring. Then, the public will be better informed about the private military debate. Also, increased accountability would allow contractors injured or killed to be medically cared for. Nevertheless, this change, though significant, will be difficult, and it will not solve the other issues this book brings to light. To that end, I recommend the following initiatives: (1) institute improved contractual control over PSCs; (2) develop a robust governmental oversight capability; (3) develop better ways to integrate outside business practices such as subcontracting into the security sector; (4) legally restrict the types of jobs for which PSCs are employed (e.g., security guards, sentries, prison guards, etc.); (5) acknowledge that PSC contractors are not civilians and adopt appropriate legal definitions and frameworks for PSC contractors; (6) develop the legal regime to "police" the companies themselves as well as their employees (formalizing the ISOA voluntary code of conduct is a step in this direction); and, (7) actively work towards eliminating the use of private force.

First, states must institute improved contractual control over PSCs. This seems like an obvious fix, and states like the United States and Great Britain have done extensive work in this area. But, changes in regulations alone are insufficient to ensure that there is adequate contractual control. Carafano (2008) and others note that the way to better improve contractual control is first achieved through the contracts themselves. They must address cost savings, of course, but the contracts must establish the scope of work in close coordination with the military and other principles' aims. Each contract, while limited in size, plays a role in the overall campaign, and these roles must be deconflicted to ensure a better unity of purpose. Here is a sample of other required inclusions. Contracts must stipulate the command and reporting relationships, clearly define both the means to achieve the contracts as well as their endstate, specify the detailed system of oversight, and include following the LOAC. The contracts must also provide the needed accountability of the personnel, to include training, equipping, and care before and during operations. They must also address the plan for the possible death or injuries of the employees, perhaps to a stipulated quality and level of care. These are only the basic ingredients. In addition to better contracts, the government must have skilled and knowledgeable personnel who let the contracts, oversee their progress and completion, report and adjudicate violations, etc.

These lead to the second recommendation: develop a robust governmental oversight capability. In order to achieve a robust oversight capability, states must invest in both knowledgeable personnel and systems to provide it. Too often in the last decade, the state has been playing catchup with the private military sector. Having MEJA and changing the wording of those who are legally covered under a NDAA is a start, but it is haphazard and reactionary. We have had years to work with the privatization phenomenon; we should have robust oversight systems.

Since it is clear that PMCs, including PSCs, are here for the foreseeable future, we need to find creative ways to integrate outside business practices such as subcontracting into the security sector. We should also harness the innovative talent of our best companies to ensure that those best business practices—the ones we know we need to adopt—are used to better the performance of the private military services in meeting the requirements of the government as the principle.

Specifically, in terms of armed contractors in the PSCs, if we believe that these kinds of services and companies are necessary, then we can limit them. We should consider legally restricting the types of jobs for which PSCs are employed, such as security guards, sentries, prison guards, and so on. While these services and job types fall short of those Erik Prince envisions, limiting the roles of PSCs helps to provide the accountability and oversight that as discussed, are so problematic and that draw the ire of so many.

Furthermore, as I argued in Chapter 5, armed contractors are not civilians, and they are not legal combatants either. We must acknowledge that private security contractors are not innocent civilians, whom it is both illegal and immoral to target, and we should adopt appropriate legal definitions and frameworks for PSC contractors. Combatantcy comes with an added privilege that civilians do not enjoy. If captured during war, combatants gain prisoner status (POW); no longer combatants, they are now not targetable. Civilians are usually not considered POW-eligible. Legally, civilians may be detained for criminal activity, but according to the LOAC, because they were not taking an active part in the hostilities, they would normally not be entitled to become POWs if captured. In wartime, professional soldiers are combatants. The question is whether the PSC contractor is a combatant as well. The US government of course maintains that contractors, including armed PSC contractors, are civilians, but they are not. PSC contractors are not currently hired to conduct offensive operations; however, the lines between offensive operations and defensive ones are often blurred. By being armed, conducting military related services, and operating in an uncertain wartime environment, and precisely because, even if they will not intentionally engage in combat, they are there, armed and ready, on the presumption that there will be combat, PSC contractors are combatants of a certain sort. It is time to recognize these moral and legal distinctions, and formulate laws that recognize and legalize their status.

While the legal uncertainty of the status of private contractors continues to be debated, the companies themselves have attempted to begin a regulation regime to voluntarily police each other and the industry. One might skeptically believe that these companies are forming groups like the ISOA and the BAPSC to merely deflect poor public perception. Yet, this may only be a positive byproduct. Companies understand that perception often equates to favorable competition for contracts. Regulating themselves is one way to ensure the market weeds out the poorly run and the more nefarious companies. As discussed, however, the market has not been so successful at regulating this complex industry, and there remains opportunities for less costly and less regulated security companies as well. A totalitarian state or a non-state actor such as a drug cartel or a group like Boko Haram might not desire a LOAC-following, state licensed PSC. Nevertheless, taking the cue from the companies themselves, I recommend incorporating some of their regulatory practices into the private military contracting law.

Finally, I think that states and the international community should actively work towards eliminating the use of private force. If states claim the monopoly on the legitimate use of force, then this force should not be private. States can, of course delegate that monopoly, for example through contracting private military companies. The contract formalizes this delegation, and the money exchanged is merely compensation for the service. But, I think that the rush to redefine Weber's words to allow for the acceptable use of PMCs and PSCs misses the larger picture. It is conceivable that wide-spread hiring of PSCs could dramatically alter how we perceive states and their functions. The problem then is not that states no longer exist;

rather, the problem is that long term reliance upon privatized force may change conceptions of statehood in the future. In addition, the struggle in Iraq to eliminate private militias, the international search for a solution to the pirates off of the Somali coast, and governments' conflicts with drug cartels demonstrates how states and international organizations are expending large amounts of energy and resources to eliminate competing private forces. Nevertheless, the hiring of PSCs reintroduces private force back into the equation. The question for all of us is whether this turn to private military and the resurgence of private force is acceptable; force becomes a commodity. I have argued that force should not be commodified, and if it is already a commodity to a degree, it seems that is both morally and pragmatically preferable to work towards limiting the degree to which force is a commodity.

This list is not complete, of course, yet implementing these recommendations will start the process of better regulating PMCs and PSCs. And, it draws both the problems and the benefits of the use of private security into the public's debate. In addition, I must share a word about how to better integrate them. How am I as a military planner and commander on the ground going to be integrated with these security companies? I think that this question is best answered by looking at four interrelated areas: accountability; communication, unity of effort; and training.

When my unit first deployed to Iraq in 2004, we were, quite frankly, surprised by both the number of PSC companies operating in our battlespace and the degree to which privatization overall, and not just logistics, had become part of how we fought. Accountability was one of the first areas that we tackled. Accountability in war is paramount to ensure there are no incidents of friendly fire and that after the battle, no one is left behind. Our units and the CPA eventually recognized this friction point, and through personal relationships—literally built through a personal census—combined with the establishment of an informal PSC working group, we were able to better establish the needed contacts to ensure our operations and the security companies' missions were deconflicted. The working group grew enough that eventually Aegis was hired to coordinate the activities of PSCs operating in Iraq.

While it was often unclear to us who worked for whom, a network of communication was established to ensure that no one would operate completely independent and outside of rescue. A "sheriff's net" was used so that anyone in need, coalition forces, PSC, PMC, DOS, and so on could call for needed assistance. Yet, these communication plans were created after the mission was underway, through trial and error, and by necessity. The best practices were adopted and shared across the theaters, but these communication plans and accountability systems must be planned for in advance by all parties, prior to the employment of military forces or PSCs.

Integrating PSCs, then, must become part of military units' tactical and operational planning. Not only must we develop systems for accountability and communication with the participating PSCs, but this planning must ensure that how the military and the security companies operate (as well as when and where) are leading towards the accomplishment of the objectives and the overall mission. Unity of effort is an essential military tenant of planning and operating. All activities must contribute to the state's mission, and during the conflict, we must ensure that all of the military and PSC actions are assessed by appropriate measures of performance and effectiveness to monitor and adjust them as needed to confirm that their effects are in fact contributing and not disrupting the campaign.

Nevertheless, planning and assessing the unity of effort, as well as developing and implementing systems of accountability and communication are critical, they are not easy tasks. The military (and others) need the requisite training. Furthermore, units need homestation training on operating with PSCs. The tasks listed above involve collective training and are focused on the headquarters. Yet, the soldiers themselves need training as well. Recall that my soldiers had no idea that there would be privatized sniper teams in support of security escorts until we met them. We used to call it "contractors on the battlefield" or "civilians on the battlefield" training. But, merely acknowledging that soldiers will be interacting with contractors is not enough. As the military has recognized the imperative for cultural training, including how to better understand and operate in different cultures, so too must the military include such training for integrating and interacting with security contractors.

Parting shots

I trust that the reader has come to appreciate the complex debate over the hiring of PSCs. Certainly, the private contractors of today have a long and colorful history in their lineage; it is in part their mercenary heritage that calls their employment into question. Nevertheless, the debate over PSCs should not be one of mere financial saving and efficiency, nor should it be solely focused on transparency and government accountability. While these are certainly important, there is, as I have argued, an ethical dimension to this debate as well. Additionally, the answer to the question, "is it morally acceptable to hire PSCs?" should be answered in the negative, all things considered. There should be a prima facie reluctance to use PSCs unless in extreme circumstances, where their use results in an overall better state of affairs.

I make this claim based upon the three discussed arguments. First, while there is no absolute prohibition against commodifying force, generally speaking, states should not hire PSCs as this practice leads to the negative consequences of force commodification. The PSC proponents' own cost-benefit calculations demonstrate that the costs, financial or otherwise, far outweigh the expected benefits of using PSCs. Force may be treated as a commodity, but it should not be.

Second, there exists a belligerent equality among combatants that does not extend to PSC contractors. As I discussed, McMahan was incorrect in arguing that MEC is false, and it is MEC that illustrates a clear distinction between PSC contractors and combatants. Even those who still think that the MEC debate is not solved, as well as those who agree with McMahan, would agree that the moral and legal status of armed security contractors has yet to be settled. While current international law categorizes PSC employees and other contractors as civilians, I have shown that this move is shortsighted and is incorrect. PSC contractors may not be combatants, but they are certainly not civilians either. A legal category for PSC contractors is needed.

Third, the hiring of PSCs affects the military as a profession. The military's expertise, corporateness, and responsibility are each affected. These in turn change civil-military relations in Western democracies, and as I have argued, alters in the ad bellum debate over deciding when to go to war. This change also affects the public debate over when states should conduct peace operations on humanitarian grounds.

Of course, as many would point out, not all of the effects of hiring PSCs are negative, and I highlighted the many benefits of their use. Moreover, I have met and worked with many of these contractors, whom I have found to be professional in their conduct and very competent in their work. Nevertheless, currently the balance sheet lies in favor of not using PSCs. Until the circumstances change the balance of consequences, maybe in cases such as the relief operations in Haiti or in Darfur, states should not hire PSCs. We should also be cautious about the calls to hire private military to fight ISIS or Boko Haram. Not only should force be considered as only one of or part of any solution, so too must the consideration of private force and security.

An opponent might interject, saying that my argument, while convincing and ought to be considered for the future employment of PSCs, does not fully address PSCs currently being used. She would be correct to an extent; we cannot turn back the privatization clock. PSCs are in use, and it would be naive to believe that they would disappear in the near future. However, my argument points out that (1) PSCs should not be used unless extreme circumstances; (2) states must understand the moral implications of future employment of PSCs; and, (3) states should institute programs of oversight now and reduce the issues of control and other negative consequences. Furthermore, states must also work towards reducing their reliance upon privatized force.

To this end, I suggested that states at a minimum ought to (1) institute improved contractual control over PSCs; (2) develop a robust governmental oversight capability; (3) develop better ways to integrate outside business practices such as sub-contracting into the security sector; (4) legally restrict the types of jobs for which PSCs are employed (e.g., security guards, sentries, prison guards, etc.); (5) acknowledge that PSC contractors are not civilians and adopt appropriate legal definitions and frameworks for PSC contractors; (6) develop the legal regime to "police" the companies themselves as well as their employees; and, (7) actively work towards eliminating the use of private force. This list is of course only partially complete, and much work needs to be done—by governments, academics, practitioners, and the public—to reduce the overall negative consequences of the current use of PSCs.

Furthermore, in this chapter I have highlighted the need for contractor accountability, care, and transparency. I have also noted the steps that the military should take to ensure that those PSCs currently being employed and the ones that may be hired in the future are better integrated with the military and political planning, operations, and endstate.

Finally, the commodification of force, the moral and legal status of security contractors, and the possible effects of private security on the military as a profession should be part of a wider public debate on whether states should use armed private security and in what circumstances.

If we do not have this debate now, then we may be too late in discovering the unforeseen, increased consequences due to the evolving use of private force. As late as 2008–2009, revelations uncovered that the CIA turned to Blackwater to help design and implement a covert program to locate, target, and perhaps assassinate Al Qaeda leadership (Ciralsky 2010; Mazzetti 2009). Although there were no reported killings, for almost three years, the CIA used Blackwater to develop this capability. When CIA Director Leon Panetta closed the program and informed Congress, Congress was evidently not previously informed of the operation. There was no public declaration, no discussion, no debate. While details of how far developed this program was remain unclear, Prince admits that his company helped locate at least one target in October 2008 (Ciralsky 2010).

Even though the CIA's hiring of Blackwater may not seem to directly affect the military as a profession, and while it does not necessarily call into question belligerent equality nor the actual commodification of force, it does demonstrate evidence of an undisclosed use of private military by a government agency, and it reveals a lack of legislative oversight, let alone public debate, further underlining why the PSC debate is so critical.

Notes

1 Hageborn notes that Middlebury College scholar Allison Stanger first coined the phrase "First Contractors' War."
2 For an in-depth exploration of the piracy issue and the growing use of private maritime security companies, see Pitney Jr. and Levin's *Private Anti-Piracy Navies* (2013) and Berube and Cullen's *Maritime Private Security: Market Responses to Piracy, Terrorism and Waterborne Security Risks in the 21st Century* (2012).
3 The rsb-group.org website includes a France24.com video news report on Russian PMCs. RSB Group is highlighted, but so is Slavanic Corps, who according to the video, was involved in the fighting in Syria. These former Russian soldiers were arrested upon return to Russia.
4 A number of public interest groups are attempting to track these contractor coasts and shed public light on their sacrifice. For example see the "Disposable army: Civilian Contractors in Iraq and Afghanistan," http://www.propublica.org/series/disposable-army.

Bibliography

Abrahamsen, Rita and Michael C. Williams. 2011. *Security Beyond the State: Private Security in International Politics*. Cambridge: Cambridge University Press.

Adams, Thomas K. 1999. "The New Mercenaries and the Privatization of Conflict." *Parameters* 29(2) (summer): 103–16.

Addington, Larry H. 1994. *The Patterns of War since the Eighteenth Century*. 2nd ed. Bloomington: Indiana University Press.

Aftergood, Stevan. 2009. "Defense Contracting in Afghanistan at Record High." Available at: www.fas.org/blog/secrecy/2009/09/afghan_contracting.html [accessed October 15, 2014].

Alexandra, Andrew, Deane-Peter Baker and Marina Caparini. 2009. *Private Military and Security Companies: Ethics, Policies and Civil-Military Relations*. New York: Routledge.

Anders, Birthe. 2014. "Private Military and Security Companies: A Review Essay." *Parameters* 44(2) (summer): 75–80.

Annan, Kofi. 1998. Ditchley Foundation Lecture XXXV. June 26. Available at: www.ditchley.co.uk/conferences/past-programme/1990-1999/1998/lecture-xxxv [accessed April 23, 2016].

Anscombe, Elizabeth. 1970. "War and Murder." In *War and Morality*, ed. Richard A. Wasserstrom, 42–53. Belmont, CA: Wadsworth Publishing.

Anscombe, G. E. M. 1981. *Ethics, Religion and Politics*. Oxford: Blackwell.

Apuzzo, Matt. 2009. "Pentagon Letter Undercuts DOJ in Blackwater Case." *Associated Press*, February 3.

Aristotle, W. D. Ross, J. O. Urmson and J. L. Ackrill. 1980. *The Nicomachean Ethics*. Oxford; New York: Oxford University Press.

Art, Robert J. and Kenneth Neal Waltz. 1999. *The Use of Force: Military Power and International Politics*. 5th ed. Lanham, MD: Rowman & Littlefield.

Ashcroft, James and Clifford Thurlow. 2006. *Making a Killing: The Explosive Story of a Hired Gun in Iraq*. London: Virgin Books.

Avant, Deborah D. 2004. "Think Again: Mercenaries." *Foreign Policy* (July/August). Available at: www.du.edu/korbel/sie/media/documents/faculty_pubs/avant_2004_foreign_policy_think_again.pdf [accessed April 23, 2016].

—. 2005. *The Market for Force: The Consequences of Privatizing Security*. Cambridge; New York: Cambridge University Press.

—. 2007. "The Emerging Market for Private Military Services and the Problems of Regulation." In *From Mercenaries to Market: The Rise and Regulation of Private Military Companies*, ed. Simon Chesterman and Chia Lehnardt, 181–95. Oxford: Oxford University Press.

Axelrod, Alan. 2014. *Mercenaries: A Guide to Private Armies and Private Military Companies*. Thousand Oaks, CA: CQ Press/Sage Publications.

Bacevich, A. J. 2008. *The Limits of Power: The End of American Exceptionalism.* 1st ed. New York: Metropolitan Books.

Bailes, Alyson, Ulrich Schneckener and Herbert Wulf. 2006. "Revisiting the State Monopoly on the Legitimate Use of Force." Geneva Centre for the Democratic Control of Armed Forces (DCAF). www.isn.ethz.ch/isn/Digital-Library/Publications/Detail/?ots591=0C54E3B3-1E9C-BE1E-2C24-A6A8C7060233&lng=en&id=30834 [accessed October 20, 2008].

Baker, Deane-Peter. 2010. *Just Warriors, Inc. Armed Contractors and the Ethics of War.* London: Continuum.

Baker, Peter, Helene Cooper and Michael R. Gordon. 2015. "Obama Looks at Adding Bases and Troops in Iraq, to Fight ISIS." *New York Times.* June 11. www.nytimes.com/2015/06/12/world/middleeast/iraq-isis-us-military-bases-martin-e-dempsey.html?_r=0 [accessed April 23, 2016].

Barber, Richard W. 1980. *The Reign of Chivalry.* New York: St. Martin's Press.

Barnes, David M. 1999. *The Problem of Intervention.* MA thesis. University of Massachusetts.

—. 2009. "Jus in Bello and the Sophisticated Utilitarian." Presented at the International Symposium on Military Ethics. San Diego, CA. Available at: http://isme.tamu.edu/ISME09/Barnes09.html [accessed April 23, 2016].

—. 2013. "Should Private Security Companies be Employed for Counterinsurgency Operations?" *Journal of Military Ethics* 12(3): 201–24, DOI: 10.1080/15027570.2013.847535.

Barry, James A. 1998. *The Sword of Justice: Ethics and Coercion in International Politics.* Westport, CT: Praeger.

Bartels, Elizabeth. 2008. "Private Sector, Public Wars: An Interview with Dr. James Jay Carafano." *Journal of International Peace Operations* 4(2): 19–24.

BBC.com. 2009. "CIA Cancels Blackwater Drone Missile-Loading Contract." December 12. Available at: http://news.bbc.co.uk/2/hi/8409358.stm [accessed April 23, 2016].

Beah, Ishmael. 2007. *A Long Way Gone: Memoirs of a Boy Soldier.* New York: Farrar, Straus and Giroux.

Beitz, Charles R. and Lawrence A. Alexander. 1985. *International Ethics.* Princeton, NJ: Princeton University Press.

Benbaji, Yitzhak. 2008a. "The Responsibility of Soldiers and the Ethics of Killing in War." *The Philosophical Quarterly* 57(229): 558–72.

—. 2008b. "The War Convention and the Moral Division of Labour." *The Philosophical Quarterly* 59: 593–617. DOI: 10.1111/j.1467-9213.2008.577.x.

—. 2008c. "A Defense of the Traditional War Convention." *Ethics* 118(3): 464–95.

Benson, Bruce L. 2007. "The Market for Force." *The Independent Review* 11(3): 451–8.

Berube, Claude. 2008. "Uncharted Waters." *Journal of International Peace Operations* 4(2): 17–18.

Berube, Claude and Patrick Cullen. 2012. *Maritime Private Security: Market Responses to Piracy, Terrorism and Waterborne Security Risks in the 21st Century.* Abingdon: Routledge.

Bicanic, Nick and Jason Bourque. 2006. *Shadow Company.* DVD. Purpose.

Birtle, Andrew J. 1998. *US Army Counterinsurgency and Contingency Operations Doctrine 1860–1941.* Washington, DC: Center of Military History.

Blackwater USA. n.d. Available at: http://blackwaterusa.com [accessed April 23, 2016].

Blackwater USA. 2005. "Blackwater Joins Hurricane Katrina Relief Effort!" Press Release. Available at: www.blackwaterusa.com/btw2005/archive/090505btw.html [accessed November 10, 2008].

Blackwater Worldwide. 2008. Press Release. 16 October. Available at: http://blackwaterme-diacenter.com/images/pdf/10-16-08%20Maritime-Release.pdf [accessed November 10, 2008].

Blizzard, Stephan M. 2004. "Increasing Reliance on Contractors on the Battlefield: How Do We Keep Them from Crossing the Line?" *Air Force Journal of Logistics* 28(1): 2–13.

Blum, William. 2004. *Killing Hope: US Military and CIA Interventions since World War II*. 2nd ed. Monroe, ME: Common Courage Press.

Bomann-Larsen, Lene. 2004. "License to Kill? The Question of Just Vs. Unjust Combatants." *Journal of Military Ethics* 3(2): 142–60.

—. 2007. *Reconstructing the Moral Equality of Soldiers*. Oslo: Acta Humaniora.

Boot, Max. 2006. "Send in the Mercenaries: Darfur Needs Someone to Stop the Bloodshed, Not More Empty U.N. Promises," *Council on Foreign Relations*. 31 May. Available at: www.cfr.org/publication/10798/send_in_the_mercenaries.html [accessed 28 October 2008].

Boutros-Ghali, Boutros. 1999. *Unvanquished: A US–UN Saga*. 1st ed. New York: Random House.

Brandt, R. B. 1972. "Utilitarianism and the Rules of War." *Philosophy and Public Affairs* 1(2): 145–65.

Brannen, Kate. 2015. "The Company Getting Rich Off the ISIS War." DailyBeast.com. August 2. Available at: www.thedailybeast.com/articles/2015/08/02/the-company-get-ting-rich-off-of-the-isis-war.html [accessed April 23, 2016].

Bremer, L. Paul and Malcolm McConnell. 2006. *My Year in Iraq: The Struggle to Build a Future of Hope*. New York: Simon & Schuster.

British Association of Private Security Companies (BAPSC). 2006. www.bapsc.org.uk/.

Brown, Chris. 2002. *Sovereignty, Rights, and Justice: International Political Theory Today*. Cambridge; Malden, MA: Polity; Blackwell Publishers.

Brown, Peter G. 2000. *Ethics, Economics and International Relations: Transparent Sovereignty in the Commonwealth of Life Edinburgh Studies in World Ethics*. Edinburgh: Edinburgh University Press.

Bruneau, Thomas. 2011. *Patriots for Profit: Contractors and the Military in US National Security*. Stanford, CA: Stanford University Press.

Burke, James. 2002. "Expertise, Jurisdiction, and Legitimacy of the Military Profession." In *The Future of the Army Profession*, ed. Don Snider and Gayle L. Watkins, 19–38. Boston: McGraw Hill.

Businesswire. 2003. "Cubic to Continue Warfighter Education Support for US Army Combined Arms Center at Fort Leavenworth; Awarded $9.8 Million with Possibility of Four Option Years." November 25. Available at: www.businesswire.com/news/home/20031125005040/en/Cubic-Continue-Warfighter-Education-Support-U.S.-Army#.Va6CMWPeI8w [accessed July 21, 2015].

Byers, Michael. 2005. *War Law: Understanding International Law and Armed Conflict*. 1st American ed. New York: Grove Press.

Calhoun, Laurie. 2001. "Killing, Letting Die, and the Alleged Necessity of Military Intervention." *Peace and Conflict Studies* 8(2): 5–21.

Callahan, David. 1997. *Unwinnable Wars: American Power and Ethnic Conflict*. New York: Hill and Wang.

Camm, Frank and Victoria A. Greenfield. 2005. *How Should the Army Use Contractors on the Battlefield?* Santa Monica, CA: RAND Corporation.

Campbell, Greg. 2004. *Blood Diamonds: Tracing the Deadly Path of the World's Most Precious Stones*. New York: Basic Books.

Carafano, James Jay. 2008. *Private Sector, Public Wars: Contractors in Combat—Afghanistan, Iraq, and Future Conflicts the Changing Face of War.* Westport, CT: Praeger.

Carmola, Kateri. 2004. "Outsourcing Security: The Ethical and Legal Consequences of Private Military Corporations." Presented at International Studies Association Annual Meeting. Montreal, Quebec, March.

—. 2007. "Ethical Borderlands in Polymorphic Wars: PMCs and the Rise of Contract Ethics." Middlebury College.

—. 2010. *Private Security Contractors in the Age of New Wars: Risk, Law & Ethics.* New York: Routledge Press.

Carstens, Roger D., Michael A. Cohen and Maria Figueroa Kupcu. 2008. *Changing the Culture of Pentagon Contracting.* Washington, DC: New America Foundation.

Ceulemans, Carl. 2008. "The Moral Equality of Combatants." *Parameters* 37(4) (winter): 99–109.

Challans, Timothy L. 2004. "Worthy and Unworthy Wars: A Kantian Critique of Just War Reasoning." Presented at International Symposium on Military Ethics. Available at: http://isme.tamu.edu/JSCOPE04/Challans04.html [accessed April 23, 2016].

—. 2007. *Awakening Warrior: Revolution in the Ethics of Warfare.* Albany, NY: State University of New York Press.

Chatterjee, Pratap. 2004. "Ex-SAS Men Cash in on Iraq Bonanza." Corpwatch.org. 9 June. Available at: www.corpwatch.org/article.php?id=11355 [accessed April 23, 2016].

Cheadle, Don and John Prendergast. 2007. *Not on Our Watch: The Mission to End Genocide in Darfur and Beyond.* New York: Hyperion.

Cheney-Peters, Scott. 2014. "Whither the Private Maritime Security Companies of South and Southeast Asia?" Center for International Maritime Security. April 16. Available at: http://cimsec.org/whither-the-pmscs/10878 [accessed September 3, 2015].

Chesterman, Simon. 2002. *Just War or Just Peace?: Humanitarian Intervention and International Law.* Oxford; New York: Oxford University Press.

Chesterman, Simon and Chia Lehnardt. 2007. *From Mercenaries to Market: The Rise and Regulation of Private Military Companies.* Oxford; New York: Oxford University Press.

Christopher, Paul. 1999, 2004. *The Ethics of War and Peace: An Introduction to Legal and Moral Issues.* 2nd/3rd ed. Upper Saddle River, NJ: Prentice Hall.

—. 2006. "Civilians Contractors: Ethical and Legal Parameters." *Journal of International Peace Operations* 2(2): 9.

Cicero, Marcus Tullius and Clinton Walker Keyes. 1928. *De Re Publica*; De Legibus Loeb Classical Library. Cambridge, MA; London: Harvard University Press; W. Heinemann.

Ciralsky, Adam. 2010. "Tycoon, Contractor, Soldier, Spy." *Vanity Fair* (Jan). Available at: www.vanityfair.com/politics/features/2010/01/blackwater-201001?printable=true [accessed April 23, 2016].

Citino, Robert Michael. 2005. *The German Way of War: From the Thirty Years' War to the Third Reich.* Modern War Studies. Lawrence, KS: University Press of Kansas.

Clark, Roger Stenson and Madeleine Sann. 1996. *The Prosecution of International Crimes.* New Brunswick, CT: Transaction Publishers.

Clark, Wesley K. 2001. *Waging Modern War: Bosnia, Kosovo, and the Future of Combat.* New York: Public Affairs.

Clausewitz, Carl von, Michael Eliot Howard and Peter Paret. 1976. *On War.* Princeton, NJ: Princeton University Press.

Coady, C. A. J. (Tony). 2008. "The Status of Combatants." In *Just and Unjust Warriors: The Moral and Legal Status of Soldiers*, ed. David Rodin and Henry Shue, 153–75. Oxford; New York: Oxford University Press.

Coates, Anthony. 2008. "Is the Independent Application of Jus in Bello the Way to Limit War?" In *Just and Unjust Warriors: The Moral and Legal Status of Soldiers*, ed. David Rodin and Henry Shue, 176–92. Oxford; New York: Oxford University Press.

Cockayne, James. 2007. "Make or Buy: Principal-Agent Theory and the Regulation of Private Military Companies." In *From Mercenaries to Market: The Rise and Regulation of Private Military Companies*, ed. Simon Chesterman and Chia Lehnardt, 196–216. Oxford: Oxford University Press.

Cohen, Marshall, Thomas Nagel, Thomas Scanlon and Richard B. Brandt. 1974. *War and Moral Responsibility*. Princeton, NJ: Princeton University Press.

"Contractor Support of US Operations in the USCENTCOM Area of Responsibility." 2009. Available at: http://psm.du.edu/media/documents/reports_and_stats/us_data/dod_quarterly_census/dod_quarterly_census_may_2009.pdf [accessed October 15, 2014].

"Contractor Support of US Operations in the USCENTCOM Area of Responsibility." 2014. Available at: http://psm.du.edu/media/documents/reports_and_stats/us_data/dod_quarterly_census/dod_quarterly_census_2014_july.pdf [accessed October 15, 2014].

"Contractor Support of US Operations in the USCENTCOM Area of Responsibility." 2015. Available at: http://psm.du.edu/media/documents/reports_and_stats/us_data/dod_quarterly_census/dod_quarterly_census_july_2015.pdf [accessed September 15, 2015].

Cook, Martin L. 2000. "Moral Foundations of Military Service." *Parameters* 30(1) (spring): 117–29.

—. 2004. *The Moral Warrior: Ethics and Service in the US Military*. Albany, NY: State University of New York Press.

—. 2008. "Revolt of the Generals: A Case Study in Professional Ethics," *Parameters* 38(1): 4–15.

CORPWATCH. 2005. "State Department List of Security Companies Doing Business in Iraq," February 15. Available at: http://corpwatch.org/article.php?id=11851 [accessed September 2, 2015].

Cotton, Sarah K., Ulrich Petersohn, Molly Dunigan, Q. Burkhart, Megan Zander-Cotugno, Edward O'Connell and Michael Webber. 2010. *Hired Guns: Views About Armed Contractors in Operation Iraqi Freedom*. Santa Monica, CA: RAND Corporation.

Covell, Charles. 1998. *Kant and the Law of Peace: A Study in the Philosophy of International Law and International Relations*. Basingstoke, UK; New York: MacMillan Press; St. Martin's Press.

Cowen, Tyler. 2007. "To Know Contractors, Know Government." *New York Times*. October 28. Available at: www.nytimes.com/2007/10/28/business/28view.html?_r=0 [accessed April 23, 2016].

Cullen, Patrick. 2008. "Private Security Head-to-Head against Pirates." *Journal of International Peace Operations* 4(3): 16–20.

d'Errico, Peter. 1996. "Corporate Personality and Human Commodification." *Rethinking Marxism* 9(2): 99–113.

Daalder, Ivo H. and Michael E. O'Hanlon. 2000. *Winning Ugly: Nato's War to Save Kosovo*. Washington, DC: Brookings Institution Press.

Danspeckgruber, Wolfgang F. and Charles Tripp. 1996. *The Iraqi Aggression against Kuwait: Strategic Lessons and Implications for Europe*. Boulder, CO: Westview Press.

Daulatzai, Anila, Catherine Lutz and Ken MacLeish. 2015. *Costs of War: Allied Combatants Killed in Iraq, March 2003–April 2015*. Watson Institute for International and Public Affairs. Available at: http://watson.brown.edu/costsofwar/costs/human/military [accessed April 23, 2016].

de Castro, L. D. 2003. "Commodification and Exploitation: Arguments in Favour of Compensated Organ Donation." *Journal of Medical Ethics* 29: 142–6.

Department of Foreign Affairs and International Trade Canada (DFAIT). *Canada and Peace Operations*. n.d. Available at: www.international.gc.ca/peace-paix/index.aspx?lang=en [accessed December 16, 2009].

Dickenson, Donna. 2008. "Property in the Body: Justice, Exploitation and Choice." Presented at the Department of Philosophy, University of Colorado, Boulder, 19 August.

Dickinson. Laura A. 2011. *Outsourcing War and Peace: Preserving Public Values in a World of Privatized Foreign Affairs*. New Haven, CT: Yale University Press.

Doswald-Beck, Louise. 2007. "Private Military Companies under International Humanitarian Law." In *From Mercenaries to Market: The Rise and Regulation of Private Military Companies*, ed. Simon Chesterman and Chia Lehnardt, 115–38. Oxford: Oxford University Press.

Drennan, Justine. 2014. "Blackwater's Erik Prince Talks Fighting Ebola, ISIS, and Bad Press." *Foreign Policy*. October 28. Available at: http://foreignpolicy.com/2014/10/28/blackwaters-erik-prince-talks-fighting-ebola-isis-and-bad-press/ [accessed April 23, 2016].

Drohan, Madelaine. 2004. *Making a Killing: How and Why Corporations Use Armed Force to Do Business*. Guilford, CT: Lyon's Press.

Dunigan, Molly. 2011. *Victory for Hire: Private Security Companies' Impact on Military Effectiveness*. Stanford, CA: Stanford Security Studies.

Dworkin, Ronald. 1977. *Taking Rights Seriously*. Cambridge: Harvard University Press.

Efflandt, Scott L. 2014. Military Professionalism and Private Military Contractors. *Parameters* 44(2): 49–60.

Elsea, Jennifer K., Moshe Schwartz and Kennon H. Nakamura. 2008. *Private Security Contractors in Iraq: Background, Legal Status, and Other Issues*. Available at: http://handle.dtic.mil/100.2/ADA470189 [accessed April 23, 2016].

Engility Corporation. 2016. Available at: www.engilitycorp.com [accessed April 23, 2016].

Ergo, Comfort and Suzanne Long. 1998. "Cases and Criteria: The UN in Iraq, Bosnia, and Somalia." In *Some Corner of a Foreign Field: Intervention and World Order*, ed. Roger Williamson, 157–65. New York: St. Martin's Press.

Erickson, Andrew and Gabe Collins. 2012. "Enter China's Security Firms." TheDiplomat. com. February 21. Available at: http://thediplomat.com/2012/02/enter-chinas-security-firms/ [accessed April 23, 2016].

Ertman, Martha M. and Joan C. Williams. 2005. "Freedom, Equality, and the Many Features of Commodification." In *Rethinking Commodification: Cases and Readings in Law and Culture*, ed. Martha M. Ertman and Joan C. Williams, 1–7. New York: New York University Press.

European Centre for Conflict Prevention. 1999. *International Fellowship of Reconciliation, State of the World Forum*. Coexistence Initiative, and Sweden. Styrelsen för internationell utveckling. Utrecht, Netherlands: European Centre for Conflict Prevention, in cooperation with the International Fellowship of Reconciliation and the Coexistence Initiative of State of the World Forum.

Executive Summary, US Naval Academy's 9th Annual McCain Conference on Ethics and Military Leadership (23–24 April). 2009. *Ethics and Military Contractors: Examining the Public-Private Partnership*. Available at: www.usna.edu/ethics/Seminars/mccain.htm [accessed April 23, 2016].

Fall, Bernard B. 1961. *Street without Joy; Indochina at War, 1946–1954*. Harrisburg, PA: Stackpole Co.

FedBizOpps.gov. 2010. "The Worldwide Protective Services (WPS) program provides comprehensive protective security services to support US Department of State operations around the world." Solicitation Number: SAQMMA10R0005-A. Available at: www.fbo.

gov/index?s=opportunity&mode=form&id=23ebfe7b75e652829f01749582467aac&
tab=core&_cview=1 [accessed April 23, 2016].

Financial Times. 2009. "Corporate Warfare." Editorial, August 24. Available at: www.ft.
com/cms/s/0/0e46624a-90f2-11de-bc99-00144feabdc0.html#axzz3RIwY1GJI [accessed
February 9, 2015].

Finkel, David. 2013. *Thank You for Your Service.* New York: Sarah Crichton Books.

Fisher, David. 1998. "Some Corner of a Foreign Field." In *Some Corner of a Foreign Field:
Intervention and World Order*, ed. Roger Williamson, 28–37. New York: St. Martin's Press,
1998.

Foot, Philippa. 1967. "The Problem of Abortion and the Doctrine of Double Effect." *Oxford
Review* 5: 5–15.

Forbes, Ian and Mark Hoffman. 1993. *Political Theory, International Relations, and the
Ethics of Intervention Southampton Studies in International Policy.* New York: St. Martin's Press in association with the Mountbatten Centre for International Studies, University of Southampton.

Forsyth, Frederick. 1974. *The Dogs of War.* New York: Viking Press.

Fox, Dov. 2008. "Paying for Particulars in People-to-Be: Commercialisation, Commodification and Commensurability in Human Reproduction." *Journal of Medical Ethics* 34(3):
162–6.

Fredland, J. Eric. 2004. "Outsourcing Military Force: A Transactions Cost Perspective on
the Role of Military Companies." *Defence and Peace Economics* 15(3): 205–19.

Friedman, Milton. 1962. *Capitalism and Freedom.* Chicago: University of Chicago Press.

Frontier Services Group. 2015. Available at: fsgroup.org [accessed April 23, 2016].

—. 2016. "Chinese Private Security Firms Go Overseas: 'Crossing the River by Feeling the
Stones.'" fsgroup.org. Available at: www.fsgroup.com/chinese-private-security-firms-
go-overseas-crossing-the-river-by-feeling-the-stones/ [accessed January 16, 2015].

Gansler, Jacques S., David J. Berteau, David M. Maddox, David R. Oliver Jr., Leon E.
Salomon and George T. Singley III. 2007. *Urgent Reform Required: Army Expeditionary
Contracting. Report of the Commission on Army Acquisition and Program Management
in Expeditionary Operations.* (The Gansler Report). Available at: www.dtic.mil/dtic/tr/
fulltext/u2/a515519.pdf [accessed April 23, 2016].

Garrett, Stephen A. 1999. *Doing Good and Doing Well: An Examination of Humanitarian
Intervention.* Westport, CT: Praeger.

Gates, Robert M. 2014. *Duty: Memoirs of a Secretary at War.* New York: Alfred A. Knopf.

Gauss, Hannah. 2014. *Warlord, Inc.—The Private Security System of Afghanistan.* Cape
Town: University of Cape Town.

Giddens, Anthony. 1985. *The Nation-State and Violence.* Berkeley, CA: University of
California Press.

Gillard, Michael. 2009. "Mad, Bad, or Just Dangerous to Know?" *The Sunday Times*,
August 16.

Glanz, James & Alissa J. Rubin. 2007. "From Errand to Fatal Shot to Hail of Fire to 17
Deaths." *New York Times*, October 3. Available at: www.nytimes.com/2007/10/03/world/
middleeast/03firefight.html [accessed August 29, 2013].

Glover, Jonathan. 2001. *Humanity: A Moral History of the Twentieth Century.* New Haven,
CT: Yale Nota Bene/Yale University Press.

Goldman, Alan. 1980. *The Moral Foundations of Professional Ethics.* Totowa, NJ:
Rowman & Littlefield.

Goldstein, Joshua S. 2003. *War and Gender: How Gender Shapes the War System and Vice
Versa.* Cambridge: Cambridge University Press.

Goldstone, Richard. 2000. *For Humanity: Reflections of a War Crimes Investigator.* The Castle Lectures in Ethics, Politics, and Economics. New Haven: Yale University Press.

Goodpaster, Andrew Jackson and Carnegie Commission on Preventing Deadly Conflict. 1996. *When Diplomacy Is Not Enough: Managing Multinational Military Intervention: A Report to the Carnegie Commission on Preventing Deadly Conflict.* Washington, DC: Carnegie Commission on Preventing Deadly Conflict.

Gordon, Michael R. and Bernard E. Trainor. 2006. *Cobra II: The Inside Story of the Invasion and Occupation of Iraq.* New York: Pantheon Books.

Göring, Hermann, Erwin C. Surrency and International Military Tribunal. 1947. *Trial of the Major War Criminals before the International Military Tribunal, Nuremberg, 14 November 1945–1 October 1946.* microform. Nuremberg, Germany: [s.n.].

Grcic, Joseph. 2000. *Ethics and Political Theory.* Lanham, MD: University Press of America.

Grossman, Dave. 1996. *On Killing: The Psychological Cost of Learning to Kill in War and Society.* 1st pbk. ed. Boston, MA: Little, Brown.

Grossman, Dave and Loren W. Christensen. 2008. *On Combat: The Psychology and Physiology of Deadly Conflict in War and in Peace.* Millstadt, IL: Warrior Science Publications.

Grotius, Hugo and Francis Willey Kelsey. 1962. *The Law of War and Peace. De Jure Belli Ac Pacis, Libri Tres.* Indianapolis, IN: Bobbs-Merrill.

Hagedorn, Ann. 2014. "Is America's Second Contractors' War Drawing Near?" *Time.* Available at: http://time.com/3222342/invisible-soldiers-iraq-contractor-war/ [accessed 15 October 15, 2014].

—. 2015. *Invisible Soldiers: How America Outsourced Our Security.* New York: Simon and Schuster.

Hammes, T. X. 2011. "Private Contractors in Conflict Zones: The Good, the Bad, and the Strategic Impact." *Joint Force Quarterly* 60(1st Quarter): 26–37.

Hancock, Curtis L. and Anthony O. Simon. 1995. *Freedom, Virtue, and the Common Good.* South Bend, IN: American Maritain Association Publications.

Hare, J. E. and Carey B. Joynt. 1982. *Ethics and International Affairs.* New York: St. Martin's Press.

Hare, R. M. 1972. "Rules of War and Moral Reasoning." *Philosophy and Public Affairs* 1(2): 166–81.

Hart, H. L. A. 1994. *The Concept of Law.* 2nd ed. Oxford; New York: Clarendon Press; Oxford University Press.

Hartle, Anthony E. 2002. "Atrocities in War: Dirty hands and Noncombatants." *Social Research* 69(4): 963–79.

—. 2004. *Moral Issues in Military Decision Making.* 2nd ed. Lawrence, KS: University Press of Kansas.

Hayner, Priscilla B. 2001. *Unspeakable Truths: Confronting State Terror and Atrocity.* New York: Routledge.

Hedahl, Marc O. 2005. "Outsourcing the Profession: A Look at Military Contractors and Their Impact on the Profession of Arms." Presented at International Symposium on Military Ethics.

—. 2009. "Blood and Blackwaters: A Call to Arms for the Profession of Arms." *Journal of Military Ethics* 8(1): 19–33.

Hegel, Georg Wilhelm Friedrich and S. W. Dyde. 1996. *Philosophy of Right.* Amherst, NY: Prometheus Books.

Heinecken, Lindy. 2009. "Discontent Within the Ranks? Officers' Attitudes Toward Military Employment and Representation—A Four-Country Comparative Study." *Armed Forces & Society* 35: 477–500.

—. 2014. "Outsourcing Public Security: The Unforeseen Consequences for the Military Profession." *Armed Forces & Society* 40: 625–46.

Hipple, Matt. 2015. *Private Military Contractors: A CIMSEC Compendium*. Center for International Maritime Security's (CIMSEC). Available at: http://cimsec.org/wp-content/uploads/2015/02/pmcw.pdf [accessed February 11, 2015].

Hobbes, Thomas. 1988. *The Leviathan*. Buffalo, NY: Prometheus Books.

Hoffmann, Stanley, Robert C. Johansen and James P. Sterba. 1996. *The Ethics and Politics of Humanitarian Intervention*. Notre Dame, IN: University of Notre Dame Press.

Holbrooke, Richard C. 1998. *To End a War*. New York: Random House.

Holmes, Richard. 1986. *Acts of War: The Behavior of Men in Battle*. 1st American ed. New York: Free Press.

Holmes, Robert L. 1989. *On War and Morality Studies in Moral, Political, and Legal Philosophy*. Princeton, NJ: Princeton University Press.

Hooker, Brad. 1990. "Rule-Consequentialism." *Mind* 99(393): 67–77.

Houreld, Katharine. 2008. "Security Firms Join Somali Piracy Fight." *Associated Press*, October 27. Available at: http://usatoday30.usatoday.com/news/world/2008-10-26-2583935117_x.htm [accessed October 27, 2008].

Human Rights First. 2008. "How to End Impunity for Private Security and Other Contractors: Blueprint for the Next Administration." Available at: www.humanrightsfirst.org/wp-content/uploads/pdf/PSC-081118-end-cont-impun-blueprint.pdf [accessed April 23, 2016].

Humanitarian Studies Unit. 2001. *Reflections on Humanitarian Action: Principles, Ethics, and Contradictions*. London; Sterling, VA: Pluto Press.

Hume, David. 1967. *A Treatise of Human Nature*. Oxford: Clarendon Press.

Hume, David and J. B. Schneewind. 1983. *An Enquiry Concerning the Principles of Morals*. Indianapolis, IN: Hackett Publishing.

Huntington, Samuel P. 1957. *The Soldier and the State; The Theory and Politics of Civil-Military Relations*. Cambridge, MA: Belknap Press of Harvard University Press.

Hurka, Thomas. 2007. "Liability and Just Cause." *Ethics and International Affairs* 21: 199–218.

—. 2008. "Proportionality and Necessity." In *War and Political Philosophy*, ed. Larry May, 127–44. Cambridge: Cambridge University Press.

—. 2009. "The Consequences of War." In *Ethics and Humanity: Themes from the Writing of Jonathan Glover*, ed. Richard Keshen N. Ann Davis and Jeff McMahan, 23–43 New York: Oxford University Press.

Ignatieff, Michael. 1986. *The Needs of Strangers*. New York: Penguin Books.

—. 1998. *The Warrior's Honor: Ethnic War and the Modern Conscience*. 1st American ed. New York: Metropolitan Books.

—. 2001. *Virtual War: Kosovo and Beyond*. New York: Picador USA.

—. 2004. *The Lesser Evil: Political Ethics in an Age of Terror*. Princeton, NJ: Princeton University Press.

Institute for Security Studies: Sierra Leone. n.d. Available at: www.iss.org.za/AF/profiles/sieraleone/SecInfo.html [accessed September 30, 2009].

International Stability Operations Association (ISOA). 2011. "ISOA Code of Conduct, Ver. 13.1." Available at: www.stability-operations.org/?page=Code [accessed September 6, 2015].

International Stability Operations Association (ISOA). 2015. "Triple Canopy Wins $47.8 M Federal Protective Services Contract." News release. August 7. Available at: www.stability-operations.org/news/245215/Triple-Canopy-Wins-47.8-M-Federal-Protective-Services-Contract.htm [accessed April 23, 2016].

Isenberg, David. 2004. "A Fistful of Contractors: The Case for a Pragmatic Assessment of Private Military Companies in Iraq." British American Security Information Council (BASIC) Research Report 2004.4. London.

—. 2007. "A Government in Search of Cover: Private Military Companies in Iraq." In *From Mercenaries to Market: The Rise and Regulation of Private Military Companies*, ed. Simon Chesterman and Chia Lehnardt, 82–93. Oxford: Oxford University Press.

—. 2008. "Dogs of War: Blackwater to the Rescue in Darfur?" August 8. Available at: www.upi.com/Top_News/Special/2008/08/08/Dogs-of-War-Blackwater-to-the-rescue-in-Darfur/UPI-26991218225853/ [accessed April 23, 2016].

—. 2009. *Shadow Force: Private Security Contractors in Iraq*. Westport, CT: Praeger.

—. 2012. "The Rise of Private Maritime Security Companies." HuffingtonPost.com. May 29. Available at: www.huffingtonpost.com/david-isenberg/private-military-contractors_b_1548523.html [accessed September 3, 2015].

—. 2015. "It's a Bird? It's a Plane! No, It's a Private Drone Contractor!" *Mint Press News*. April 10. Available at: www.mintpressnews.com/MyMPN/its-a-bird-its-a-plane-no-its-a-private-drone-contractor/ [accessed April 26, 2015].

Jackson, Frank. 1991. "Decision-Theoretic Consequentialism and the Nearest and Dearest Objection." *Ethics* 101(3): 461–82.

Jäger, Thomas and Gerhard Kümmel. 2007. *Private Military and Security Companies: Chances, Problems, Pitfalls and Prospects*. Wiesbaden, Germany: VS Verlag.

Janowitz, Morris. 1960. *The Professional Soldier, a Social and Political Portrait*. Glencoe, IL: Free Press.

Johnson, James Turner. 1999. *Morality and Contemporary Warfare*. New Haven, CT: Yale University Press.

Jones, Colin. 2012. "Our Silent Partners: Private Security Contractors in Iraq." Small Wars Journal. May 17. Available at: http://smallwarsjournal.com/jrnl/art/our-silent-partners-private-security-contractors-in-iraq [accessed August 23, 2015].

Joseph, Miranda. 2005. "The Multivalent Commodity: On the Supplementarity of Value and Values." In *Rethinking Commodification: Cases and Readings in Law and Culture*, ed. Martha M. Ertman and Joan C. Williams, 383–401. New York: New York University Press.

Judd, Terri. 2009. "Steroids, Drink and Paranoia: The Murky World of the Private Security Contractor." *The Independent*. 1 September.

Kagan, Frederick W. 2006. *Finding the Target: The Transformation of American Military Policy*. New York: Encounter Books.

Kaldor, Mary. 2006. *New and Old Wars*. 2nd ed. Cambridge; Malden, MA: Polity.

Kanet, Roger E. 1998. *Resolving Regional Conflicts*. Urbana, IL: University of Illinois Press.

Kant, Immanuel and Ted Humphrey. 1983. *Perpetual Peace, and Other Essays on Politics, History, and Morals*. Indianapolis, IN: Hackett Publishing.

Kant, Immanuel, W. Hastie and Edwin D. Mead. 1914. *Eternal Peace, and Other International Essays*. Boston, MA: The World Peace Foundation.

Keegan, John. 1976. *The Face of Battle*. London: J. Cape.

—. 1999. *The First World War*. 1st American ed. New York: A. Knopf.

Keim, Willard D. 2000. *Ethics, Morality, and International Affairs*. Lanham, MD: University Press of America.

Kennedy, Robert G. 2008. "Is Just War Theory Obsolete?" Presented at the International Symposium on Military Ethics. San Diego, CA.

Kidwell, Deborah C. 2005. *Public War, Private Fight? The United States and Private Military Companies, Global War on Terrorism*. Global War on Terrorism Occasional Paper 12. Fort Leavenworth, KS: Combat Studies Institute Press.

Kinsey, Christopher. 2006. *Corporate Soldiers and International Security: The Rise of Private Military Companies*. New York: Routledge.

—. 2009. *Private Contractors and the Reconstruction of Iraq: Transforming Military Logistics*. London: Routledge.

Kinsey, Christopher and Malcolm Hugh Patterson. 2012. *Contractors and War: the Transformation of US Expeditionary Operations*. Stanford, CA: Stanford University Press.

Klare, Michael T. 2002. *Resource Wars: The New Landscape of Global Conflict*. New York: Henry Holt.

Klein, Kenneth H. and Joseph C. Kunkel. 1990. *In the Interest of Peace: A Spectrum of Philosophical Views*. Wakefield, NH: Longwood Academic.

Krahmann, Elke. 2007. "Transnational States in Search of Support: Private Military Companies and Security Sector Reform." In *From Mercenaries to Market: The Rise and Regulation of Private Military Companies*, ed. Simon Chesterman and Chia Lehnardt, 94–112. Oxford: Oxford University Press.

—. 2010. *States, Citizens and the Privatisation of Security*. Cambridge: Cambridge University Press.

Kutz, Christopher. 2008. "Fearful Symmetry." In *Just and Unjust Warriors: The Moral and Legal Status of Soldiers*, ed. David Rodin and Henry Shue, 69–86. Oxford; New York: Oxford University Press.

Lake, Eli. 2014. "Contractors Ready to Cash in on ISIS War." The Daily Beast. September 13. Available at: www.thedailybeast.com/articles/2014/09/13/contractors-ready-to-cash-in-on-isis-war.html [accessed April 23, 2016].

Lamothe, Dan. 2014. "Let Contractors Fight the Islamic State, Blackwater Founder Erik Prince Says." *Washington Post*. October 9. Available at:www.washingtonpost.com/news/checkpoint/wp/2014/10/09/let-contractors-fight-the-islamic-state-blackwater-founder-erik-prince-says [accessed April 23, 2016].

Lang, Anthony F. 2002. *Agency and Ethics: The Politics of Military Intervention*. Albany, NY: State University of New York Press.

Langfitt, Frank. 2006. "Private Military Firm Pitches Its Services in Darfur." All Things Considered: National Public Radio.

Lanning, Michael Lee. 2005. *Mercenaries: Soldiers of Fortune, from Ancient Greece to Today's Private Military Companies*. New York: Presidio Press.

Latham Jr., William C. 2009. "Not My Job: Contracting vs. Professionalism in the United States Army." *Military Review* 89(2): 40–9.

Lavelle, B. M. 1989. "Epikouroi in Thucydides." *American Journal of Philology* 110(1): 36–9.

Lawyers Committee for Human Rights. 1997. *Prosecuting Genocide in Rwanda*. Available at: www.unwatch.com/rwanda.html [accessed April 23, 2016].

Leander, Anna. 2006. *Eroding State Authority? Private Military Companies and the Legitimate Use of Force*. Rome: Rubbettino.

—. 2007. "Regulating the Role of Private Military Companies in Shaping Security and Politics." In *From Mercenaries to Market: The Rise and Regulation of Private Military Companies*, ed. Simon Chesterman and Chia Lehnardt, 49–64. Oxford: Oxford University Press.

Lederach, John Paul. 1995. *Preparing for Peace: Conflict Transformation across Cultures*. Syracuse, NY: Syracuse University Press.

—. 1997. *Building Peace: Sustainable Reconciliation in Divided Societies*. Washington, DC: United States Institute of Peace Press.

Lehnardt, Chia. 2007. "Private Military Companies and State Responsibility." In *From Mercenaries to Market: The Rise and Regulation of Private Military Companies*, ed. Simon Chesterman and Chia Lehnardt, 139–57. Oxford: Oxford University Press.

Lichtenberg, Judith. 2008. "How to Judge Soldiers Whose Cause Is Unjust." In *Just and Unjust Warriors*, ed. David Rodin and Henry Shue, 112–30. New York: Oxford University Press.

Lim, Aaron. 2014. "Privatising the Syrian war could save lives." VOXY.co.nz. February 20. Available at: www.voxy.co.nz/politics/privatising-syrian-war-could-save-lives/2523/182104 [accessed August 23, 2015].

Locke, John and C. B. Macpherson. 1980. *Second Treatise of Government*. Indianapolis, IN: Hackett Publishing.

Logan, Lara. 2007. *Blackwater.* 60 Minutes, CBS Broadcasting.

Lucas, Ann. 2005. "The Currency of Sex: Prostitution, Law, and Commodification." In *Rethinking Commodification: Cases and Readings in Law and Culture*, ed. Martha M. Ertman and Joan C. Williams, 248–70. New York: New York University Press.

Luttrell, Marcus and Patrick Robinson. 2007. *Lone Survivor: The Eyewitness Account of Operation Redwing and the Lost Heroes of Seal Team 10*. 1st ed. New York: Little, Brown.

McChrystal, Stanley. 2013. "Lincoln's Call to Service—and Ours." *The Wall Street Journal*. May 29. Available at: www.wsj.com/articles/SB100014241278873248098045785112206132991 86 [accessed February 9, 2016].

McCullough, David G. 2005. *1776*. New York: Simon & Schuster.

McFate, Sean. 2015. "Reining in Soldiers of Fortune." *New York Times*. April 18. Available at: www.nytimes.com/2015/04/18/opinion/reining-in-soldiers-of-fortune.html [accessed April 23, 2016].

——. 2016. "Return of the Mercenaries." Aeon. Available at: https://aeon.co/essays/what-does-the-return-of-mercenary-armies-mean-for-the-world [accessed January 27, 2016].

McGlen, Nancy E. and Meredith Reid Sarkees. 1993. *Women in Foreign Policy: The Insiders*. New York: Routledge.

Machairas, Dimitrios. 2014. "The Ethical Implications of the Use of Private Military Force: Regulatable or Irreconcilable?" *Journal of Military Ethics* 13(1): 49–69.

Machiavelli, Niccolò. 1520; 2003. *Art of War*. Chicago: University of Chicago Press.

Machiavelli, Niccolò and David Wootton. 1523; 1995. *The Prince*. Indianapolis, IN: Hackett Publishing.

MacIntyre, Alasdair C. 1984. *After Virtue: A Study in Moral Theory*. 2nd ed. Notre Dame, IN: University of Notre Dame Press.

McIntyre, Angela and Taya Weiss. 2007. "Weak Governments in Search of Strength: Africa's Experience of Mercenaries and Private Military Companies." In *From Mercenaries to Market: The Rise and Regulation of Private Military Companies*, ed. Simon Chesterman and Chia Lehnardt, 67–81. Oxford: Oxford University Press.

McLeary, Paul. 2015a. "US Looking for Contractors to Help in Iraq." DefenseNews.com. March 9. Available at: www.defensenews.com/story/defense/2015/03/09/us-private-contractor-iraq-isis/24654439/ [accessed April 23, 2016].

——. 2015b. "US-Trained Syrian Rebels Don't Yet Pose a Real Threat to the Islamic State." *Foreign Policy*. August 5. Available at: http://foreignpolicy.com/2015/08/07/u-s-trained-syrian-rebels-dont-yet-pose-a-real-threat-to-the-islamic-state/ [accessed April 23, 2016].

McMahan, Jeff. 1994. "Self-Defense and the Problem of the Innocent Attacker." *Ethics* 104(2): 252–90.

——. 2002. *The Ethics of Killing: Problems at the Margins of Life*. Oxford; New York: Oxford University Press.

——. 2004. "The Ethics of Killing in War." *Ethics* 114(4) Symposium on Terrorism, War, and Justice: 693–733.

——. 2006a. "Killing in War: A Reply to Walzer." *Philosophia* 34: 47–51.

—. 2006b "Liability and Collective Identity: A Reply to Walzer." *Philosophia* 34: 13–17.

—. 2006c. "On the Moral Equality of Combatants." *Journal of Political Philosophy* 14(4): 377–93.

—. 2007. "Collectivist Defenses of the Moral Equality of Combatants." *Journal of Military Ethics* 6(1): 50–9.

—. 2008. "The Morality of War and the Law of War." In *Just and Unjust Warriors: The Moral and Legal Status of Soldiers*, ed. David Rodin and Henry Shue, 19–43. Oxford; New York: Oxford University Press.

—. 2009. *Killing in War*. Oxford; New York: Clarendon Press; Oxford University Press.

Mamdani, Mahmood. 2000. *Beyond Rights Talk and Culture Talk: Comparative Essays on the Politics of Rights and Culture*. New York: St. Martin's Press.

Mansoor, Peter R. 2008. *Baghdad at Sunrise: A Brigade Commander's War in Iraq*. New Haven, CT: Yale University Press.

Mapel, David R. 1998. "Coerced Moral Agents? Individual Responsibility for Military Service." *Journal of Political Philosophy* 6(2): 18.

—. 2007. "The Right of National Defense." *International Studies Perspectives* 8: 1–15.

Marble-Barranca, Susan. 2009. "Unbecomming Conduct: Legal and Ethical Issues of Private Contractors in Military Situations." Presented at the International Society for Military Ethics. University of San Diego.

Margolick, David. 2007. "The Night of the Generals," *Vanity Fair*. April 2007. Available at: www.vanityfair.com/politics/features/2007/04/iraqgenerals200704 [accessed December 13, 2009].

Marshall, S. L. A. 1947; 2000. *Men against Fire: the Problem of Battle Command in Future War*. Washington, DC: Combat Forces Press.

Marx, Karl and T. B. Bottomore. 1964. *Early Writings*. New York: McGraw-Hill.

Marx, Karl and Friedrich Engels. 1967. *Capital: A Critique of Political Economy*. 3 vols. New York: International Publishers.

Marx, Karl, Friedrich Engels, Samuel Moore and F. Scott Fitzgerald. 1937. *Manifesto of the Communist Party*. New York: International Publishers.

Matlary, Janne Haaland and Øyvind Østerud. 2007. *Denationalisation of Defence: Convergence and Diversity*. Aldershot, UK; Burlington, VT: Ashgate.

Maurer, Peter and the United Nations General Assembly. 2008. "Montreux Document on Pertinent International Legal Obligations and Good Practices for States Related to Operations of Private Military and Security Companies During Armed Conflict." 6 October.

May, Larry. 2008. *War: Essays in Political Philosophy*. Cambridge: Cambrige University Press.

May, Larry, Eric Rovie and Steve Viner. 2006. *The Morality of War: Classical and Contemporary Readings*. 1st ed. Upper Saddle River, NJ: Pearson Education.

Mayer, Jane. 2009. "The Predator War." *The New Yorker*. 26 October. Available at: www.newyorker.com/reporting/2009/10/26/091026fa_fact_mayer [accessed April 23, 2016].

Mazzetti, Mark. 2009. "CIA Sought Blackwater's Help to Kill Jihadists." *New York Times*. August 20. Available at: www.nytimes.com/2009/08/20/us/20intel.html [accessed April 23, 2016].

Meredith, Martin. 1999. *Coming to Terms: South Africa's Search for Truth*. 1st ed. New York: Public Affairs.

Mill, John Stuart. 1859. *A Few Words on Non-Intervention*. Available at www.libertarian.co.uk/lapubs/forep/forep008.pdf [accessed June 7, 2016.

—. 1993. *On Liberty and Utilitarianism*. New York: Bantam Books.

Mill, John Stuart, and Roger Crisp. 1998. *Utilitarianism*. Oxford; New York: Oxford University Press.

Miller, J. Joseph. 2004. "Jus Ad Bellum and an Officer's Moral Obligations: Invincible Ignorance, the Constitution, and Iraq." *Social Theory and Practice* 30: 457–84.

Miller, T. Christian. 2009. "Sometimes it's Not Your War but You Sacrifice Anyway." *The Washington Post.* August 16. Available at: www.washingtonpost.com/wp-dyn/content/article/2009/08/14/AR2009081401665.html [accessed August 20, 2009].

Millett, Allan Reed. 1975. *The General: Robert L. Bullard and Officership in the United States Army, 1881–1925 Contributions in Military History.* Westport, CT: Greenwood Press.

—. 1977. *Military Professionalism and Officership in America.* Mershon Center Briefing Paper Number Two. Columbus, OH: Mershon Center.

Millett, Allan Reed, and Peter Maslowski. 1984. *For the Common Defense: A Military History of the United States of America.* New York; London: Free Press; Collier Macmillan.

Møller, Bjørn. 2005. "Privatization of Conflict: Security and War." Working paper no 2005/2. Danish Institute for International Studies.

Morgenthau, Hans J. and Council on Foreign Relations. 1969. *A New Foreign Policy for the United States.* New York: Praeger.

Morgenthau, Hans J. and Kenneth W. Thompson. 1993. *Politics among Nations: The Struggle for Power and Peace.* New York: McGraw-Hill.

Morris, Harvey. 2008. "Activists turn to Blackwater for Darfur help." *The Financial Times.* June 18. Available at: http://us.ft.com/ftgateway/superpage.ft?news_id=fto06182008210 8315688 [accessed April 23, 2016].

Morton, Joe D. 2005. "Loss of Six Blackwater Security Consulting Employees." In Statement to Employees of the Bureau of Diplomatic Security, ed. Acting Assistant Secretary for Diplomatic Security and Director for the Office of Foreign Missions. April 22. Available at: www.state.gov/m/ds/rls/rm/45056.htm [accessed January 16, 2009].

Moskos, Charles C. 1979. "From Institution to Occupation: Trends in Military Organization." In *War, Morality, and the Military Profession*, ed. Malham M. Wakin, 219–30. Boulder, CO: Westview Press.

MSNBC.com. 2004. "CIA Lacked Trained Interrogators." May 12. Available at: www.msnbc.msn.com/id/4964993/ [accessed December 13, 2009].

Musallam, Musallam Ali. 1996. *The Iraqi Invasion of Kuwait: Saddam Hussein, His State and International Power Politics.* London; New York: British Academic Press.

Nagel, Thomas. 1979. *Mortal Questions.* Cambridge; New York: Cambridge University Press.

Nissenbaum, Dion. 2014 "Private Group Sought to Arm Syrian Rebels." *The Wall Street Journal.* May 18. Available at: www.wsj.com/articles/SB1000142405270230465530457954963032455177 4?cb=logged0.6805906049843308 [accessed April 23, 2016].

Norcross, Alastair. 1997a. "Comparing Harms: Headaches and Human Lives." *Philosophy and Public Affairs* 26(2): 135–67.

—. 1997b. "Consequentialism and Commitment." *Pacific Philosophical Quarterly* 78: 380–403.

—. 1997c. "Good and Bad Actions." *Philosophical Review* 106(1): 1–34.

—. 1999. "Intending and Foreseeing Death: Potholes on the Road to Hell." *Southwest Philosophy Review* 15: 115–23.

Nossiter, Adam. 2015. "Mercenaries Join Nigeria's Military Campaign against Boko Haram." *New York Times.* March 12. Available at: www.nytimes.com/2015/03/13/world/africa/nigerias-fight-against-boko-haram-gets-help-from-south-african-mercenaries.html [accessed April 23, 2016].

Nozick, Robert. 1974. *Anarchy, State, and Utopia.* New York: Basic Books.

O'Brien, Kevin A. 2007. "What Should and What Should Not Be Regulated?" In *From Mercenaries to Market: The Rise and Regulation of Private Military Companies*, ed. Simon Chesterman and Chia Lehnardt, 29–48. Oxford: Oxford University Press.

O'Brien, William Vincent. 1979. *US Military Intervention: Law and Morality*. Beverly Hills, CA: Sage Publications.

—. 1981. *The Conduct of Just and Limited War*. New York: Praeger.

—. 1991. *Law and Morality in Israel's War with the PLO*. New York: Routledge.

O'Hanlon, Michael E. 1997. *Saving Lives with Force: Military Criteria for Humanitarian Intervention*. Washington, DC: Brookings Institution Press.

Oren, Michael B. 2007. *Power, Faith, and Fantasy: America in the Middle East, 1776 to the Present*. New York: W. W. Norton.

Orend, Brian. 2000a. *Michael Walzer on War and Justice*. Cardiff, UK: University of Wales Press.

—. 2000b. *War and International Justice: A Kantian Perspective*. Waterloo, OT: Wilfrid Laurier University Press.

—. 2006/2013. *The Morality of War*. Peterborough, OT; Orchard Park, NY: Broadview Press.

—. 2007. "The Rules of War." *Ethics and International Affairs* 21(4): 471–76.

Packer, George. 2005. *The Assassins' Gate: America in Iraq*. 1st ed. New York: Farrar, Straus and Giroux.

Palmer, R. R. 1986. "Fredrick the Great, Guibert, Bülow: From Dynastic to National War." In *Makers of Modern Strategy: From Machiavelli to the Nuclear Age*, ed. Peter Paret, Gordon Alexander Craig and Felix Gilbert, 91–119. Princeton, NJ: Princeton University Press.

Paret, Peter, Gordon Alexander Craig and Felix Gilbert. 1986. *Makers of Modern Strategy: From Machiavelli to the Nuclear Age*. Princeton, NJ: Princeton University Press.

Paterson, Andrew. 2007. *Iraq's Guns for Hire*. National Geographic Explorer.

Pattison, James. 2008. "Just War Theory and the Privatization of Military Force." *Ethics and International Affairs* 22(2): 143–62.

—. 2010a. "Deeper Objections to The Privatisation of Military Force." *Journal of Political Philosophy* 18(3): 425–47.

—. 2010b. "Outsourcing the Responsibility to Protect: Humanitarian Intervention and Private Military and Security Companies." *International Theory* 2(1): 1–31.

—. 2012. "The Legitimacy of the Military, Private Military, and Security Companies, and Just War Theory." *European Journal of Political Theory* 11(2): 131–54.

—. 2014. *The Morality of Private War: The Challenge of Private Military and Security Companies*. Oxford: Oxford University Press.

Peceny, Mark. 1999. *Democracy at the Point of Bayonets*. University Park, PA: Pennsylvania State University Press.

Pelton, Robert Young. 2006. *Licensed to Kill: Hired Guns in the War on Terror*. New York: Crown Publishers.

Percy, Sarah. 2006. *Regulating the Private Security Industry*. London: Routledge.

—. 2007a. "Morality and Regulation." In *From Mercenaries to Market: The Rise and Regulation of Private Military Companies*, ed. Simon Chesterman and Chia Lehnardt, 11–28. Oxford: Oxford University Press.

—. 2007b. *Mercenaries: The History of a Norm in International Relations*. Oxford; New York: Oxford University Press.

Perlo–Freeman, S. and E. Sköns. 2008. "The Private Military Services Industry." SIPRI Insight 001.

Perry, David. 2012. "Blackwater vs. Bin Laden: The Private Sector's Role in American Counterterrorism." *Comparative Strategy* 31(1): 41–55.

Persico, Joseph E. 1994. *Nuremberg: Infamy on Trial*. New York: Viking.

Petersohn, Ulrich. 2008. "Outsourcing the Big Stick: The Consequences of Using Private Military Companies." Working Paper 08-0129. Cambridge, MA: Weatherhead Center for International Affairs, Harvard.

Peterson, Laura. 2002. "Privatizing Combat, the New World Order," *The Center for Public Integrity*. Available at: http://projects.publicintegrity.org/bow/report.aspx?aid=148 [accessed April 16, 2009].

Pfaff, Tony. 2000. "Peacekeeping and the Just War Tradition." Carlisle, PA: Strategic Studies Institute.

Phillips, Robert L. and Duane L. Cady. 1996. *Humanitarian Intervention: Just War Vs. Pacifism Point/Counterpoint*. Lanham, MD: Rowman & Littlefield.

Pincus, Walter. 2009a. "Military Weighs Private Security on Front Lines." *The Washington Post*, July 26.

——. 2009b. "Up to 56,000 More Contractors Likely For Afghanistan, Congressional Agency Says." *The Washington Post*. December 16. Available at: www.washingtonpost.com/wp-dyn/content/article/2009/12/15/AR2009121504850.html [accessed December 18, 2009].

Pitney Jr., John J. and John-Clark Levin. 2013. *Private Anti-Piracy Navies: How Warships for Hire Are Changing Maritime Security*. New York: Lexington Books.

Plato and G. M. A. Grube. 1974. *Plato's Republic*. Indianapolis, IN: Hackett Publishing.

Pocock, J. G. A. 1975. *The Machiavellian Moment: Florentine Political Thought and the Atlantic Republican Tradition*. Princeton: Princeton University Press.

Porteus, Liza. 2005. "How Do You Like Your Contractor Money?" Fox News.com. June 30. Available at: www.foxnews.com/story/0,2933,161261,00.html [accessed April 23, 2016].

Posner, Richard A. 2005. "Community and Conscription." In *Rethinking Commodification: Cases and Readings in Law and Culture*, ed. Martha M. Ertman and Joan C. Williams, 128–32. New York: New York University Press.

Prince, Erik. 2014. *Civilian Warriors: The Inside Story of Blackwater and the Unsung Heroes of the War on Terror*. New York: Penguin.

Private Military Companies: Options for Regulation 2001–2002 (Green Paper). 2002. February 12. HCP 577. London: The Stationery Office.

Project on Government Oversight (POGO). 2009. *Pogo Letter to Secretary of State Hillary Clinton Regarding US Embassy in Kabul*. September 1. Available at: www.pogo.org/pogo-files/letters/contract-oversight/co-gp-20090901.html [accessed October 10, 2012].

ProPublica. n.d. "Disposable army: Civilian Contractors in Iraq and Afghanistan." Available at: www.propublica.org/series/disposable-army [accessed April 23, 2016].

Radin, Margaret Jane. 1996. *Contested Commodities*. Cambridge, MA: Harvard University Press.

Radin, Margaret Jane and Madhavi Sundar. 2005. "The Subject and Object of Commodification." In *Rethinking Commodification: Cases and Readings in Law and Culture*, ed. Martha M. Ertman and Joan C. Williams, 8–33. New York: New York University Press.

Railton, Peter. 1984. "Alienation, Consequentialism, and the Demands of Morality." *Philosophy and Public Affairs* 13(2): 134–71.

——. 1988. "Alienation, Consequentialism and Commitment." In *Consequentialism and Its Critics*, ed. Samuel Scheffler, vi, 294. Oxford; New York: Oxford University Press.

Ramsbotham, Oliver. 1998. "Humanitarian Intervention: the Contemporary Debate." In *Some Corner of a Foreign Field: Intervention and World Order*, ed. Roger Williamson, 62–72. New York: St. Martin's Press.

Ramsbotham, Oliver, and Tom Woodhouse. 1996. *Humanitarian Intervention in Contemporary Conflict: A Reconceptualization*. Cambridge, UK; Cambridge, MA: Polity Press.

—. 1999. *Encyclopedia of International Peacekeeping Operations*. Santa Barbara, CA: ABC-CLIO.

Ramsey, Paul. 1983. *The Just War: Force and Political Responsibility*. Lanham, MD: University Press of America.

Rawls, John. 1971. *A Theory of Justice*. Cambridge, MA: Belknap Press of Harvard University Press.

—. 1999. *The Law of Peoples; with, the Idea of Public Reason Revisited*. Cambridge, MA: Harvard University Press.

Razack, Sherene. 2004. *Dark Threats and White Knights: The Somalia Affair, Peacekeeping, and the New Imperialism*. Toronto, ON; Buffalo, NY: University of Toronto Press.

Records, Heather. 2014. "US Army Preparing to Deploy Division Headquarters to Iraq." UPI.com. September 23. Available at: www.upi.com/Top_News/US/2014/09/23/US-Army-preparing-to-deploy-division-headquarters-to-Iraq/9641411524155/ [accessed April 27, 2015].

Regan, Richard J. 1996. *Just War: Principles and Cases*. Washington, DC: Catholic University of America Press.

Reichberg, Gregory M. 2008. "Just War and Regular War: Competing Paradigms." In *Just and Unjust Warriors: The Moral and Legal Status of Soldiers*, ed. David Rodin and Henry Shue, 193–213. Oxford; New York: Oxford University Press.

Reisman, W. Michael and Chris T. Antoniou. 1994. *The Laws of War: A Comprehensive Collection of Primary Documents on International Laws Governing Armed Conflict*. New York: Vintage Books.

Reuter, Christoph. 2015. "Syria's Mercenaries: The Afghans Fighting Assad's War." *Der Spiegel*. May 11. Available at: www.spiegel.de/international/world/afghan-mercenaries-fighting-for-assad-and-stuck-in-syria-a-1032869.html [accessed April 23, 2016].

Reuters.com. 2015. "US Contractors Hurt in Bulgarian Blast were Testing Arms for Supply to Syrian Opposition." June 9. Available at: www.reuters.com/article/2015/06/09/us-bulgaria-blast-us-idUSKBN0OP12520150609 [accessed August 23, 2015].

Ricks, Thomas E. 2006. *Fiasco: The American Military Adventure in Iraq*. New York: Penguin Press.

—. 2009. *The Gamble: General David Petraeus and the American Military Adventure in Iraq, 2006–2008*. New York: Penguin Press.

Risen, James. 2012. "After Benghazi Attack, Private Security Hovers as an Issue." *New York Times*, October 12. Available at: www.nytimes.com/2012/10/13/world/africa/private-security-hovers-as-issue-after-embassy-attack-in-benghazi-libya.html?pagewanted=all&_r=0 [accessed February 11, 2015].

Roberts, Adam. 2006. *The Wonga Coup: The British Mercenary Plot to Seize Oil in Africa*. London: Profile Books.

—. 2008. "The Principle of Equal Application of the Laws of War." In *Just and Unjust Warriors: The Moral and Legal Status of Soldiers*, ed. David Rodin and Henry Shue, 226–54. Oxford; New York: Oxford University Press.

Rocheleau, Judy. 2010. "Combatant Responsibility for Fighting in Unjust Wars," *Social Philosophy Today* 26: 93–106.

Rochester, Christopher M. 2007. *A Private Alternative to a Standing United Nations Peacekeeping Force*. Washington, DC: Peace Operations Institute.

Rodin, David. 2002. *War and Self-Defense*. Oxford; New York: Clarendon Press; Oxford University Press.

—. 2007. "The Liability of Ordinary Soldiers for Crimes of Aggression." *Washington University Global Studies Law Review* 6: 591–607.

—. 2008. "The Moral Inequality of Soldiers: Why Jus in Bello Asymmetry is Half Right." In *Just and Unjust Warriors: The Moral and Legal Status of Soldiers*, ed. David Rodin and Henry Shue, 44–68. Oxford; New York: Oxford University Press.

Rodin, David and Henry Shue. 2008. *Just and Unjust Warriors: The Moral and Legal Status of Soldiers*. Oxford; New York: Oxford University Press.

Rogin, Josh. 2014. "Exclusive: Russian 'Blackwater' Takes Over Ukraine Airport." Daily-Beast.com. February 28. Available at: www.thedailybeast.com/articles/2014/02/28/exclusive-russian-blackwater-takes-over-ukraine-airport.html [accessed April 23, 2016].

Rose, Carol M. 2005. "Whither Commodification?" In *Rethinking Commodification: Cases and Readings in Law and Culture*, ed. Martha M. Ertman and Joan C. Williams, 402–27. New York: New York University Press.

Rosen, Armin. 2014. "Erik Prince Is Right: Private Contractors Will Probably Join the Fight against ISIS." Business Insider.com. October 10. Available at: www.businessinsider.com/why-eric-prince-is-right-about-isis-2014-10 [accessed April 23, 2016].

Rosenthal, Joel H. and Carnegie Council on Ethics & International Affairs. 1999. *Ethics & International Affairs: A Reader*. 2nd ed. Washington, DC: Georgetown University Press.

Roston, Aram. 2015. "Blackwater Founder Erik Prince Pitched Private Fighting Force To Nigeria For War Against Boko Haram." BuzzFeed.com. March 17. Available at: www.buzzfeed.com/aramroston/blackwater-founder-erik-prince-pitched-private-fighting-forc [accessed August 23, 2015].

Rotberg, Robert I. 1999. *Creating Peace in Sri Lanka: Civil War and Reconciliation*. Cambridge, MA: World Peace Foundation and the Belfer Center for Science and International Affairs Brookings Institution Press.

Rothman, Jay. 1992. *From Confrontation to Cooperation: Resolving Ethnic and Regional Conflict Violence, Cooperation, Peace*. Newbury Park, CA: Sage.

Rothwell, Donald R. 2004. "Legal Opinion on the Status of Non-Combatants and Contractors under International Humanitarian Law and Australian Law." Australian Strategic Policy Institute. Available at: www.aspi.org.au/pdf/ASPIlegalopinion_contractors.pdf [accessed November 3, 2009].

Rousseau, Jean-Jacques. 1893. *The Social Contract*. New York: G. P. Putnam's Sons.

RSB-Group (Russian Security Systems). n.d. Available at: rsb-group.org [accessed April 23, 2016].

RT.com. 2014. "Bill to Allow Private Military Contractors Submitted to Russian Parliament." October 22. Available at: http://rt.com/politics/198176-russian-private-military-contractors/ [accessed February 11, 2015].

Rubin, Alissa J. and Rod Nordland. 2014. "Sunni Militants Advance Toward Large Iraqi Dam." *New York Times*. June 25. Available at: www.nytimes.com/2014/06/26/world/middleeast/isis-iraq.html?_r=0 [accessed April 23, 2016].

Ryan, Cheyney. 2008. "Moral Equality, Victimhood, and the Sovereignty Symmetry Problem." In *Just and Unjust Warriors: The Moral and Legal Status of Soldiers*, ed. David Rodin and Henry Shue, 131–52. Oxford; New York: Oxford University Press.

Sallyport Global. 2016. Available at: www.sallyportglobal.com [accessed April 23, 2016].

Salopek, Paul. 2007. "Desperation Drives Kin of Four Abducted Mercenaries to Speak Out. An Exodus of Highly Paid Guns Alarms, Embarrasses Pretoria." *Los Angeles Times*, October 17, A10.

Sandel, Michael J. 1998. "What Money Can't Buy: The Moral Limits of Markets." Oxford: The Tanner Lectures on Human Values, May.

—. 2012. *What Money Can't Buy: The Moral Limits of Markets*. New York: Farrar, Straus and Giroux.

Sandline International. 2004. Available at: http://www.sandline.com [accessed April 23, 2016].

Sarin, O. L. and L. S. Dvoretsky. 1993. *The Afghan Syndrome: The Soviet Union's Vietnam*. Novato, CA: Presidio.

Scahill, Jeremy. 2007. *Blackwater: The Rise of the World's Most Powerful Mercenary Army*. New York: Nation Books.

—. 2009. "The Secret US War in Pakistan." The Nation.com. November. Available at: www.thenation.com/article/secret-us-war-pakistan/ [accessed August 17, 2015].

Scharf, Michael P. 1997. *Balkan Justice: The Story Behind the First International War Crimes Trial since Nuremberg*. Durham, NC: Carolina Academic Press.

Schaub, Gary and Volker Franke. 2009. "Contractors as Military Professionals?" *Parameters* 39(4): 88–104.

Scheffler, Samuel. 1988. *Consequentialism and Its Critics*. Oxford; New York: Oxford University Press.

Schmitt, Eric and Ben Hubard. 2015. "US Revamping Rebel Force Fighting ISIS in Syria." *New York Times*. September 6. Available at: www.nytimes.com/2015/09/07/world/middleeast/us-to-revamp-training-program-to-fight-isis.html?ref=middleeast [accessed April 23, 2016].

Schoonhoven, Richard. 2003. "Invincible Ignorance and the Moral Equality of . . . Lawyers?" Presented at the Joint Services Conference on Professional Ethics, January 2003. Available at: http://isme.tamu.edu/JSCOPE03/Shoonhoven03.html [accessed April 23, 2016].

Schreirer, Fred and Marina Caparini. 2005. *Privatizing Security: Law, Practice and Governance of Private Military and Security Companies*. Geneva: Geneva Center for the Democratic Control of Armed Forces (DCAF), March.

Schumacher, Gerald. 2006. *A Bloody Business: America's War Zone Contractors and the Occupation of Iraq*. Grand Rapids, MI; St. Paul, MN: Zenith; MBI Publishing.

Schütte, Robert. 2014. *Civilian Protection in Armed Conflicts: Evolution, Challenges and Implementation*. Wiesbaden, Germany: Springer.

Schwartz, Moshe. 2009. "Department of Defense Contractors in Iraq and Afghanistan: Background and Analysis." In CRS Report to Congress: Congressional Research Service, August 19.

—. 2010. "Department of Defense Contractors in Iraq and Afghanistan: Background and Analysis." In CRS Report to Congress: Congressional Research Service, July 2.

—. 2013. "Department of Defense's Use of Contractors to Support Military Operations: Background, Analysis, and Issues for Congress." May 17. Available at: www.fas.org/sgp/crs/natsec/R43074.pdf [accessed August 23, 2015].

Shalom, Stephen R. 2011. "Killing in War and Moral Equality." *Journal of Moral Philosophy* 8(4): 495–512.

Shevory, Kristina. 2014. "A Former Mercenary Proposes a Pentagon Makeover." *Newsweek*. December 2. Available at: www.newsweek.com/2014/12/12/former-mercenary-proposes-pentagon-makeover-288327.html [accessed April 23, 2016].

Shorrock, Tim. 2014. "Who Profits From our New War? Inside NSA and Private Contractors' Secret Plans." Salon.com. September 24. Available at: www.salon.com/2014/09/24/heres_who_profits_from_our_new_war_inside_nsa_and_an_army_of_private_contractors_plans/# [accessed April 23, 2016].

—. 2016. "Contractor Kidnappings and the Perils of Privatized War." The Nation.com. Available at: www.thenation.com/article/contractor-kidnappings-and-the-perils-of-privatized-war/ [accessed February 2, 2016].

Shue, Henry. 1996. *Basic Rights: Subsistence, Affluence, and US Foreign Policy*. 2nd ed. Princeton, NJ: Princeton University Press.

——. 2008. "Do We Need a 'Morality of War'?" In *Just and Unjust Warriors: The Moral and Legal Status of Soldiers*, ed. David Rodin and Henry Shue, 87–111. Oxford; New York: Oxford University Press.

Sidgwick, Henry. 1919. *The Elements of Politics*. 4th ed. London; New York: Macmillan Kraus Reprint.

——. 1981. *The Methods of Ethics*. 7th ed. Indianapolis, IN: Hackett Publishing.

Silverstein, Ken and Daniel Burton-Rose. 2000. *Private Warriors*. New York: Verso.

Singer, P. W. 2003. *Corporate Warriors: The Rise of the Privatized Military Industry*. Cornell Studies in Security Affairs. Ithaca, NY: Cornell University Press.

——. 2005. "Outsourcing War." Foreign Affairs. March 1. Available at: www.foreignaffairs. com/articles/2005-03-01/outsourcing-war [accessed April 23, 2016].

——. 2006. *Children at War*. New York: Pantheon Books.

——. 2007a. *Can't Win with 'Em, Can't Go to War without 'Em: Private Military Contractors and Counterinsurgency*. Washington, DC: Brookings Institution Press.

——. 2007b. "We Can't Fight the War without the Company—but We Won't Win with It on Our Payroll." *The Washington Post*. October 5. Available at: www.washingtonpost.com/ wp-dyn/content/discussion/2007/10/05/DI2007100501642_pf.html [accessed April 23, 2016].

——. 2008. "Corporate Warriors: The Privatized Military and Iraq." Carnegie Council on Ethics and International Affairs. Transcript of remarks. Available at: www.carnegiecoun-cil.org/studio/multimedia/20051201/index.html [accessed October 20, 2008].

Sizemore, Bill. 2005. "US: Private Security Company Creates Stir in New Orleans." The Virginian Pilot. September 15. Available at: www.corpwatch.org/article.php?id=12634 [accessed November 10, 2008].

Smith, Michael. 2006. *Killer Elite: The Inside Story of America's Most Secret Special Operations Team*. London: Weidenfeld & Nicolson.

Smith, Michael A. 1995. *Human Dignity and the Common Good in the Aristotelian-Tho-mistic Tradition*. Lewiston, NY: Mellen University Press.

Snider, Don. 2002. *The Future of the Army Profession*. Boston: McGraw Hill.

——. 2009. *The Army's Ethic Suffers under Its Retired Generals*. Carlisle, PA: Strategic Studies Institute.

Snider, Don and Gayle L. Watkins. 2000. "The Future of Army Professionalism: A Need for Renewal and Redefinition." *Parameters* 30(3): 5–20.

Snider, Don, John A. Nagel and Tony Pfaff. 1999. *Army Professionalism, the Military Ethic, and Officership in the 21st Century*. Carlisle, PA: Strategic Studies Institute.

Source Watch. 2007. "Blackwater USA – Sourcewatch." Available at: www.sourcewatch. org/index.php?title=Blackwater_USA [accessed April 23, 2016].

Spearin, Christopher. 2011. "UN Peacekeeping and the International Private Military and Security Industry." *International Peacekeeping* 17(2): 196–209.

——. 2014. "Special Operations Forces and Private Security Companies." *Parameters* 44(2): 61–73.

Spicer, Tim. 1999. *An Unorthodox Soldier: Peace and War and the Sandline Affair: An Autobiography*. Edinburgh: Mainstream.

Spielmann, Peter James. 2013. "Global Security Industry Expanding," *The Associated Press, Stars and Stripes*, 11, November 6.

Spurlock, Timothy (ed.) 2001. *West Point's Perspectives on Officership: Selected Readings*. New York: Alliance Press.

Statman, Daniel. 2006. "Supreme Emergencies Revisited." *Ethics* 117(1): 58–79.

Steingart, Gabor. 2009. "Merchants of Death: Memo Reveals Details of Blackwater Targeted Killings Program," *Spiegel*. August 24. Available at: www.spiegel.de/international/world/0,1518,644571,00.html [accessed April 23, 2016].

Stepanova, Ekaterina. "Trends in Armed Conflicts: One-Sided Violence against Civilians." Stockholm International Peace Research Institute. Available at: www.sipri.org/yearbook/2009/02 [accessed April 23, 2016].

Sterling, Scott A. 2004. "Moral Equality among Soldiers." In *Peace and War*. Center for Christian Ethics: Baylor University, 69–73.

Stoker, Donald J. 2008. *Military Advising and Assistance: From Mercenaries to Privatization, 1815–2007*. London; New York: Routledge.

Stroble, James A. 1996. "Theories of Noncombatant Immunity." In *The Ethics of War and the Uses of War*. Phd diss., University of Hawaii.

Struerchler, Nikolas. 2008. "The Swiss Initiative Comes Alive." *Journal of International Peace Operations* 4(3): 9–12.

Sun tzu, Ralph Sawyer, Mei-chün Sawyer and Pin Sun. 1996. *The Complete Art of War History and Warfare*. Boulder, CO: Westview Press.

Surowiecki, James. 2004. "Army, Inc." *The New Yorker*, January 12.

Talty, Stephan. 2007. *Empire of Blue Water: Captain Morgan's Great Pirate Army, the Epic Battle for the Americas, and the Catastrophe That Ended the Outlaws' Bloody Reign*. New York: Crown Publishers.

Tesón, Fernando R. 1998. *A Philosophy of International Law: New Perspectives on Law, Culture, and Society*. Boulder, CO: Westview Press.

—. 2005. *Humanitarian Intervention: An Inquiry into Law and Morality*. 3rd ed. Ardsley, NY: Transnational Publishers.

The Guardian. 2004. "Seven Killed in Baghdad Suicide Blast." May 6. Available at: www.theguardian.com/world/2004/may/06/iraq.usa [accessed April 27, 2015].

Thomas and Dominicans. English Province. 1947. *Summa Theologica*. 1st complete American ed. New York: Benziger Bros.

Thompson, Loren. 2013. "As Budget Cuts Bite, Defense Contractors' Fortunes Diverge." *Forbes*. April 24. Available at: www.forbes.com/sites/lorenthompson/2013/04/24/pentagon-service-providers-face-tougher-market-tighter-margins/#45d823fd1c9a [accessed February 8, 2016].

Thomson, Janice E. 1994. *Mercenaries, Pirates, and Sovereigns: State-Building and Extraterritorial Violence in Early Modern Europe*. Princeton, NJ: Princeton University Press.

Thomson, Judith Jarvis. 1990. *The Realm of Rights*. Cambridge, MA: Harvard University Press.

—. 1991. "Self-Defense." *Philosophy and Public Affairs* 20(4): 283–310.

Thucydides, Robert B. Strassler and Richard Crawley. 1996. *The Landmark Thucydides: A Comprehensive Guide to the Peloponnesian War*. New York: Free Press.

Tickner, J. Ann. 1992. *Gender in International Relations: Feminist Perspectives on Achieving Global Security*. New Directions in World Politics. New York: Columbia University Press.

Tilly, Charles, Gabriel Ardant and Social Science Research Council (US). 1975. *Committee on Comparative Politics. The Formation of National States in Western Europe Studies in Political Development*. Princeton, NJ: Princeton University Press.

Toffler, Alvin and Heidi. 1995. *War and Anti-War: Making Sense of Today's Global Chaos*. London: Warner Books.

Toner, James H. 1993. "Teaching Military Ethics." *Military Review* 73: 33–40.

Tonkin, Hannah. 2011. *State Control over Private Military and Security Companies in Armed Conflict*. Cambridge: Cambridge University Press.

Triple Canopy. n.d. Available at: www.triplecanopy.com/home/ [accessed April 23, 2016].

Trundle, Matthew. 2004. *Greek Mercenaries: From the Late Archaic to Alexander*. London; New York: Routledge.

United Nations (UN). 1949. General Assembly. *Universal Declaration of Human Rights*. Washington, DC: US GPO.

—. 1989. *International Convention against the Recruitment, Use, Financing and Training of Mercenaries*. A/RES/44/34. 72nd plenary meeting. December 4. (UN Mercenary Convention). Available at: www.un.org/documents/ga/res/44/a44r034.htm [accessed April 23, 2016].

—. 2008. "UN List of Peacekeeping Missions.". Available at: www.un.org/Depts/dpko/list/list.pdf [accessed April 23, 2016].

—. 2015. "UN List of Peacekeeping Missions." Available at: www.un.org/en/peacekeeping/resources/statistics/factsheet.shtml [accessed July 21, 2015].

United States Commission on Wartime Contracting in Iraq and Afghanistan. 2009. *Lessons from the Inspectors General Improving Wartime Contracting*. A Hearing of the Commission on Wartime Contracting, February 2009. Washington, DC: Commission on Wartime Contracting in Iraq and Afghanistan. Available at: http://purl.access.gpo.gov/GPO/LPS113665 [accessed April 23, 2016].

—. 2012. "Transforming Wartime Contracting: Controlling Costs, Reducing Risks." Available at: www.whs.mil/library/Reports/CWC_FinalReport-lowres.pdf [accessed August 23, 2015].

United States Congress House. Committee on Oversight and Government Reform. 2006. Subcommittee on National Security Emerging Threats and International Relations. *Private Security Firms Standards, Cooperation, and Coordination on the Battlefield*. Hearing before the Subcommittee on National Security, Emerging Threats, and International Relations of the Committee on Government Reform, House of Representatives, One Hundred Ninth Congress, Second Session, June 13, 2006. Washington: US GPO.

—. 2007a. *Blackwater USA: Private Military Contractor Activity in Iraq and Afghanistan*: Hearing before the Committee on Oversight and Government Reform, House of Representatives, One Hundred Tenth Congress, First Session, October 2, 2007. Washington, DC: US GPO.

—. 2007b. *Iraqi Reconstruction: Reliance on Private Military Contractors and Status Report*. Hearing before the Committee on Oversight and Government Reform. House of Representatives, One Hundred Tenth Congress, First Session, February 7, 2007. Washington, DC: US GPO.

—. 2007c. Committee on Homeland Security and Governmental Affairs. *Accountability in Government Contracting Act of 2007*. Report of the Committee on Homeland Security and Governmental Affairs, United States Senate to Accompany S. 680, to Ensure Proper Oversight and Accountability in Federal Contracting, and for Other Purposes. Washington, DC: US GPO.

—. 2007d. John Warner National Defense Authorization Act for Fiscal Year 2007. Public Law 109-364, October 17, 2006. 109th Congress, H.R. 5122.

—. 2008a. *Employment Practices of Blackwater Worldwide*. Memorandum from the Committee on Oversight and Government Reform, House of Representatives, One Hundred Tenth Congress, March 10, 2008. Washington, DC: US GPO.

—. 2008b. Committee on Armed Services. *Accountability During Contingency Operations Preventing and Fighting Corruption in Contracting and Establishing and Maintaining Appropriate Controls on Materiel*. Washington, DC: US GPO. Available at: http://purl.access.gpo.gov/GPO/LPS109371 [accessed April 23, 2016].

—. 2009. Committee on Armed Services. *Inquiry into the Treatment of Detainees in US Custody.* Washington, DC: US GPO. Available at: http://purl.access.gpo.gov/GPO/LPS112554 [accessed April 23, 2016].

United States Department of Defense, Inspector General (DOD IG). 2015. "Contingency Contracting: A Framework for Reform–2015 Update." Washington, DC: US Department of Defense. March 31, 2015. Available at: www.dodig.mil/pubs/documents/DODIG-2015-101.pdf [accessed April 23, 2016].

United States Department of State. 2009. "Security Companies Doing Business in Iraq." Available at: www.micahmwhite.com/why-war-archive/security-companies-doing-business-in-iraq-2005 [accessed April 23, 2016].

United States Department of the Army. 1956a. *The Law of Land Warfare.* Washington, DC: US GPO.

—. 1956b. *Treaties Governing Land Warfare.* Washington, DC: US GPO.

—. 1999. *Contracting Support on the Battlefield*, FM 4-100.2. Washington, DC: US Department of the Army.

—. 2003a. *Peace Ops: Multi-Service Tactics, Techniques, and Procedures for Conducting Peace Operations*, Field Manual 3-07.31. Washington, DC: US Department of the Army.

—. 2003b. *Stability Operations and Support Operations*, Field Manual 3-07. Washington, DC: US Department of the Army.

—. 2003c. *Contractors on the Battlefield*, FM 3-100.21. Washington, DC: US Department of the Army.

—. 2006. *Counterinsurgency*, Field Manual 3-24. Washington, DC: US Department of the Army.

—. 2008. *Operations.* Field Manual 3.0. Washington, DC: US Department of the Army.

—. 2011. *Unified Land Operations*, ADP 3-0. Washington, DC: US Department of the Army.

—. n.d. US Army Peacekeeping and Stability Operations Institute (PKSOI). Available at: https://pksoi.army.mil/ [accessed April 23, 2016].

United States Federal Acquisition Regulation (FAR). 2015. "Subpart 7.5-Inherently Governmental Functions." Available at: www.acquisition.gov/?q=/browse/far/7 [accessed August 23, 2015].

United States Government Accountability Office (GAO). 2006. Draft Report. *Defense Logistics: Changes to Stryker Vehicle Maintenance Support Should Identify Strategies for Addressing Implementation Challenges.* September 5, 2006. GAO-06-928R.

—. 2008. *Rebuilding Iraq DOD and State Department Have Improved Oversight and Coordination of Private Security Contractors in Iraq, but Further Actions Are Needed to Sustain Improvements*: Report to Congressional Committees. [Rev. ed.] Washington, DC: US Government Accountability Office. Available at: http://purl.access.gpo.gov/GPO/LPS99996 [accessed April 23, 2016].

United States US Joint Chiefs of Staff. 1999. *Joint Tactics, Techniques, and Procedures for Peace Operations.* Washington, DC: US Joint Chiefs of Staff.

United States US Joint Forces Command. 1999. *Armed Private Security Contractor Project Handbook (Draft).* Washington, DC: Joint Futures Lab, US Joint Forces Command, April.

Uttley, Matthew. 2005. "Contractors on Deployed Military Operations: United Kingdom Policy and Doctrine." Carlisle, PA: Strategic Studies Institute.

Van Creveld, Martin. 2008. *The Culture of War.* New York: Presidio Press/Ballantine Books.

Vanden Brook, Tom, Ken Dilanian and Ray Locker. 2009. "Retired Military Officers Cash in as Well-Paid Consultants." *USA Today.* November 18. Available at: www.usatoday.com/news/military/2009-11-17-military-mentors_N.htm [accessed December 13, 2009].

Verkuil, Paul R. 2007. *Outsourcing Sovereignty: Why Privatization of Government Functions Threatens Democracy and What We Can Do About It.* New York: Cambridge University Press.

Vincent, R. J. 1974. *Nonintervention and International Order.* Princeton, NJ: Princeton University Press.

Vitoria, Francisco de, Anthony Pagden and Jeremy Lawrance. 1991. *Political Writings.* Cambridge; New York: Cambridge University Press.

Volcker, Paul A. 2003. *Urgent Business for America: Revitalizing the Federal Government for the 21st Century.* Report of the National Commission on the Public Service. Washington, DC: Brookings Institution Press. Available at: www.brookings.edu/~/media/research/files/reports/2003/1/01governance/01governance.pdf [accessed June 7, 2016].

Von Lipsey, Roderick K. 1997. *Breaking the Cycle.* New York: St. Martin's Press.

Wakin, Malham M. 1979. *War, Morality, and the Military Profession.* Boulder, CO: Westview Press.

Walker, David M., United States. Congress. House. Committee on Appropriations. Subcommittee on Department of Defense and United States Government Accountability Office. 2007. *Stabilizing and Rebuilding Iraq Conditions in Iraq Are Conducive to Fraud, Waste, and Abuse.* Testimony before the Subcommittee on Defense, Committee on Appropriations, House of Representatives Testimony GAO-07-525 T. Washington, DC: US Government Accountability Office. Available at: http://purl.access.gpo.gov/GPO/LPS83037 [accessed April 23, 2016].

Walzer, Michael. 1977, 2000, 2006a. *Just and Unjust Wars: A Moral Argument with Historical Illustrations.* New York: Basic Books.

—. 1980. "The Moral Standing of States: A Response to Four Critics." *Philosophy and Public Affairs* 9(3): 209–29.

—. 1983. *Spheres of Justice: A Defense of Pluralism and Equality.* New York: Basic Books.

—. 1994. *Thick and Thin: Moral Argument at Home and Abroad.* Notre Dame, IN: University of Notre Dame Press.

—. 2004. *Arguing About War.* New Haven, CT: Yale University Press.

—. 2006b. "Terrorism and Just War." *Philosophia* 34: 3–12.

—. 2008. "Mercenary Impulse." *The New Republic* 238(4): 20–1.

Warner, Margaret. 2009. "Security Contractors Under Scrutiny at US Embassy in Kabul." PBS.org. September 2. Available at: www.pbs.org/newshour/bb/asia-july-dec09-embassy_09-01/ [accessed April 23, 2016].

Wasserstrom, Richard A. 1970. *War and Morality.* Belmont, CA: Wadsworth.

Weaver, Teri. 2007. "Tiny Base Assimilates into Japanese Town." *Stars and Stripes.* October 8. Available at: www.stripes.com/news/tiny-base-assimilates-into-japanese-town-1.69654 [accessed April 23, 2016].

Weber, Max. 1926. "Politik Als Beruf." [Politics as Vocation]. On-line translation available at http://anthropos-lab.net/wp/wp-content/uploads/2011/12/Weber-Politics-as-a-Vocation.pdf [accessed April 23, 2016].

Weber, Max, A. M. Henderson and Talcott Parsons. 1947. *The Theory of Social and Economic Organization.* 1st American ed. New York: Oxford University Press.

Weinberger, Sharon. 2007. "Facing Backlash, Blackwater Has a New Business Pitch: Peacekeeping." *Wired.* December 18. Available at: http://archive.wired.com/politics/security/news/2007/12/blackwater [accessed April 23, 2016].

Weiner, Rebecca Ulam. 2006. "As the International Community Dithers Over Darfur, Private Military Companies Say they've Got What It Takes to Stop the Carnage, If Only

Someone Would Hire Them." *The Boston Globe.* April 23. Available at: www.boston. com/news/globe/ideas/articles/2006/04/23/peace_corp/ [accessed April 23, 2016].

Weiss, Michael. 2013. "The Case of the Keystone Cossacks." *Foreign Policy.* November 21. Available at: http://foreignpolicy.com/2013/11/21/the-case-of-the-keystone-cossacks/ [accessed April 23, 2016].

Weiss, Thomas George. 1999. *Military-Civilian Interactions: Intervening in Humanitarian Crises.* Lanham, MD: Rowman & Littlefield.

Wells, Donald A. 1996. *An Encyclopedia of War and Ethics.* Westport, CT: Greenwood Press.

Wertheimer, Roger. 2007. "Reconsidering Combatant Moral Equality." *Journal of Military Ethics* 6(1): 60–74.

West, Francis J. 2005. *No True Glory: A Frontline Account of the Battle for Fallujah.* New York: Bantam Books.

Whalen, Richard J. 2006. "Revolt of the Generals." *The Nation.* October 16. Available at: www.thenation.com/article/revolt-generals/ [accessed April 23, 2016].

Wheeler, Nicholas J. 2000. *Saving Strangers: Humanitarian Intervention in International Society.* Oxford; New York: Oxford University Press.

Whitworth, Sandra. 2004. *Men, Militarism, and UN Peacekeeping: A Gendered Analysis.* Boulder, CO: Lynne Rienner.

Williams, Bernard Arthur Owen. 1981. *Moral Luck: Philosophical Papers, 1973–1980.* Cambridge; New York: Cambridge University Press.

Williamson, Roger. 1998. *Some Corner of a Foreign Field: Intervention and World Order.* New York: St. Martin's Press.

Wines, Michael. 2004. "An African Foul-Up, With an Intriguing Cast of Britons." *New York Times*, September 3. Available at: www.nytimes.com/2004/09/03/world/africa/an-african-foulup-with-an-intriguing-cast-of-britons.html [accessed August 23, 2015].

Witte, Griff. 2005. "Private Security Contractors Head to Gulf." *The Washington Post.* September 8. A14.

Woodhouse, Tom and Oliver Ramsbotham. 2000. *Peacekeeping and Conflict Resolution.* London; Portland, OR: Frank Cass.

Xenophon, Rex Warner and George Cawkwell. 1978. *A History of My Times* (Hellenica) Harmondsworth, UK; New York: Penguin Books.

Zarate, Juan Carlos. 1998. "The Emergence of a New Dog of War: Private International Security Companies, International Law, and the New World Disorder." *Stanford Journal of International Law* 34: 75–162.

Zenko, Micha. 2015. "The New Unknown Soldiers of Afghanistan and Iraq." *Foreign Policy.* May 29. Available at: http://foreignpolicy.com/2015/05/29/the-new-unknown-soldiers-of-afghanistan-and-iraq/ [accessed April 23, 2016].

Zohar, Noam J. 2004. "Innocence and Complex Threats: Upholding the War Ethic and the Condemnation of Terrorism." *Ethics* 114(4) Symposium on Terrorism, War, and Justice: 734–51.

Zupan, Daniel. 2006. "The Moral (in)Equality of Combatants." Presented at the Joint Services Conference on Professional Ethics. Available at: http://isme.tamu.edu/ JSCOPE06/Zupan06.html [accessed April 23, 2016].

—. 2008. "A Presumption of the Moral Equality of Combatants: A Citizen-Soldier's Perspective." In *Just and Unjust Warriors: The Moral and Legal Status of Soldiers*, ed. David Rodin and Henry Shue, 214–25. Oxford; New York: Oxford University Press.

Index